大董中国意境菜　七十二物候

DaDong
Artistic Conception of Chinese Cuisine
THE 72 PHENOLOGICAL PERIODS
2023

大董 著　/　谢安冰 编　/　〔澳〕杰夫·龙（Geoff Lung）　大董 摄影

電子工業出版社
Publishing House of Electronics Industry
北京·BEIJING

East's charm, in art's arm.

The true allure of the East is encapsulated in its artistic sensibilities. It's a grace that's reminiscent of the Wei and Jin dynasties, a luminescence as vivid as the moon in the Tang and Song eras, and as dramatic as the crashing waves and swirling snow.

In my forty-year culinary journey, I've traversed countless miles alone, immersed in the realms of gastronomy, photography, and calligraphy along the serene rivers and lush fields. My quest for beauty has been unyielding, a journey that's taken me from the northernmost to the southernmost corners of the country. I've witnessed and tasted the essence of the land, merging culinary prowess with the artistic spirit of the landscapes. This has allowed me to translate the sophistication of literati beyond the mere dimensions of color, aroma, and flavor. Over a decade ago, I proposed the concept of "Dadong's Chinese Artistic Cuisine" and have since brought it to life. Using ingredients as my paint, techniques as my rhythm, presentation as my palette, and seasons as my muse, I've transformed picturesque vistas into culinary masterpieces. While preserving the aesthetic traditions of Eastern cuisine, I've also integrated a modern interpretation of beauty.

Two years ago, I spearheaded the creation of a novel concept, Gastro Esthetics at DaDong, a holistic dining experience designed to elevate the sensory journey at Dadong Restaurant and redefine the art of contemporary dining. The space, adorned with abstract, minimalist curves, evokes a slowly unfurling scroll depicting a panorama of rivers and mountains. The muted color scheme, dominated by shades of grey and accented with splashes of red, relies heavily on the interplay of light and shadow, creating a rich and atmospheric dining environment that exudes sophistication and grandeur. This concept necessitated a rethinking of the relationship between space, food, music, and tableware, each element bearing its unique character but all unified by the essence of beauty. This multi-dimensional immersion in 'beauty' captures fleeting moments in time and offers an experience that blends Eastern and Western aesthetics. It also aligns with my vision for Dadong's Chinese Artistic Cuisine in this new era.

"Dadong's Chinese Artistic Cuisine" introduces something novel nearly every year. Each passing moment signifies growth and evolution. We unearth and explore the bounty nurtured by mountains, rivers, and seas through the ever-changing seasons, and transform them into the 'beauty' that graces our plates. This is the heart of Dadong's Chinese Artistic Cuisine. From the four seasons to the twenty-four solar terms, and further refined into seventy-two phenological periods, we adhere to the philosophy of "every moment is a masterpiece" and "eating in season". The ancient wisdom of timing and geography is beautifully encapsulated in the jade plates and exquisite dishes that grace the Dadong dining table.

As the seasons ebb and flow, Dadong's Chinese Artistic Cuisine aims to distill the essence of each season through culinary techniques, transcending time and space to create an aesthetic and sensory experience that appeals to all. It's akin to Picasso's transition from academic aesthetics to abstract beauty, or Bada Shanren (Zhu Da) of the East, who achieved simplicity through complexity, with minimal strokes yet imbued with deep artistic charm. This aligns with Dadong's pursuit of "individual beauty, shared beauty, and the grand beauty of Dadong".

After assimilating international perspectives and years of market refinement, Dadong's Chinese Artistic Cuisine's twenty-four solar terms table and seventy-two phenological dishes will become cultural symbols, reflecting the distinctive and profound meaning of Eastern culinary aesthetics. I hope this book serves as an inspiration to us all.

Every encounter has a prelude, and kindred spirits resonate on the same frequency. As we each journey on our paths, we're united by the wind that sweeps across a thousand miles. Whether our meeting is early or late, it crystallizes into an eternal moment in an instant.

DaDong

东方之美，美在意境

魏晋风骨、唐宋明月，转瞬千年，大浪淘沙，惊涛拍岸，卷起千堆雪。

从业四十多年，一人千里，清川菱行，美食、摄影、书法……我一直坚定地行走在寻找和发现美的路上，大江南北，亲历印证，将厨艺与山水意境结合，诠释色香味之外的文人雅致。提出"大董中国意境菜"理论并付诸实践十多年，以食材为笔墨、技法为韵律、盘碟为画布、时令为表达，把眼睛看见的道道风景，变成餐桌上的丝缕意境。传承东方餐饮美学的同时，亦加入了这个时代对美的理解。

两年前，我带领团队打造了一个全新的"美·大董"整体空间概念，旨在升级大董餐厅的五感体验，重新定义当代用餐艺术。用抽象的极简弧线重组空间，挥翰雕琢千里江山画卷。以低饱和度灰为主导的节制用色，仅以红色点缀，更多地依靠光影来塑造朦胧和丰富层次感的用餐环境，是一种高级感和大写意，在这个概念里，我们重新梳理了空间、美食、音乐、器皿的关系，各有特性，但无一不以美为其用，打造出一场关于"美"的立体沉浸式体验，为的是留住时空里的吉光片羽，体验充满东西方美学交融的意境，亦符合我对大董中国意境菜在新时期的表达和期许。

《大董中国意境菜》几乎每年都会推陈出新，岁月中的分分秒秒，皆是万物生长的进程，发现并挖掘山川河海在时节交替中蕴养的珍物，并将其化为皿中之"美"，正是大董中国意境菜的核心表达。从四季到二十四节气，再到这本探究细化的七十二物候，皆秉承了"四六时中、皆为绝色"和"不时不食"的理念，古人对天时地理的细腻捕捉，一一对应在大董餐桌上的玉盘珍馐。

时节轮转，大董中国意境菜尝试以料理技巧留住时令之味，突破了时间与空间，最终形成一场受众共同意会的味觉美学体验。正如毕加索从初期学院派审美逐步向抽象之美的蜕变，亦如东方的八大山人，大繁若简，寥寥几笔，却深藏神韵，这与大董追求的"各美其美，美人之美，美美与共，大董大美"不谋而合。

大董中国意境菜经过国际视野的融入和多年的市场洗练，二十四节气餐桌和七十二物候美食，将成为东方料理美学中极具辨识度和深长意味的文化符号。愿以此书你我共勉。

相遇有伏笔，知音是同频。
于道各努力，千里自同风。
相逢话早晚，一瞬即永恒。

contents 目录

立春
Beginning of Spring

01/72 物候 东风解冻

香椿炸酱面 012
Traditional Beijing Zhajiang Noodles With Soybean Paste, Pork And Toon

没包完的饺子 015
Unfinished Dumpling

02/72 物候 蛰虫始振

油浸橄榄菜拌蚕豆 016
Broad Bean With Preserved Olive Cabbage

樱桃花鹅肝 018
Pomegranate Foie Gras

03/72 物候 鱼陟负冰

小葱拌豆腐 021
Tofu Spring Onion

香糟春塘片 022
Fish Fillets In Fermented Rice Wine

雨水
Rain Water

04/72 物候 獭祭鱼

无油水煮比目鱼配干式熟成牛肉 028
Spicy Boiled Sole With Dry Aged Sirloin

黑海盐奶油豆蓉汤 031
Sweet Pea Soup With Black Sea Salt

05/72 物候 候雁北

番茄脆菇沙拉 032
Crispy Mushroom In Tomato

油醋汁无渣芹菜 035
Crispy Celery With Vinaigrette

06/72 物候 草木萌动

豌豆小时候 036
Tender Peas

糟煨冬笋 038
Braised Bamboo Shoot With Fermented Glutinous Rice

惊蛰
Insects Awaken

07/72 物候 桃始华

春笋锅巴炒翅 045
Fried Shark Fin With Rice Crust

春草黑鱼子蘑菇沙拉 046
Marinated Mushroon With Caviar

08/72 物候 仓庚鸣

鲜花椒芽炝加蚌 049
Sliced Canadian Geoduck Clam With Sichuan Pepper Sprout

青蒜焗黄鱼 050
Green Garlic With Yellow Croaker

09/72 物候 鹰化为鸠

油菜花沙拉 052
Canola Flower Salad

韭菜苔炒望潮 054
Fried Baby Octopus With Chives

春分
Spring Equinox

10/72 物候 玄鸟至

火灼烟熏雪花牛肉配油菜花 060
Wagyu With Canola Flower

油醋汁童子菜 063
Sweet And Sour Crown Daisy

11/72 物候 雷乃发声

老糟蒸岱衢族大黄鱼卷 064
Steamed Yellow Croaker Roll In Fermented Glutinous Rice Fragrance

红漾番茄带子汤 066
Red Tomato Soup With Scallop

12/72 物候 始电

雪菜蛏子龙须贡面 069
Dragon Whiskers Noodles With Potherb Mustard And Razor Clam

荠菜酱蚕豆羊肚菌 070
Fried Morel And Broad Bean With Shepherd's Purse Pesto

清明
Pure Brightness

13/72 物候 桐始华

五花肉烧粉皮桃花泛 076
Braised Pork Belly And Green Bean Starch Sheet

问政山笋荠菜包子 078
Steam Bamboo Shoot And Shepherd's Purse Bun

14/72 物候 田鼠化为鴽

樱桃萝卜 081
Shredded Cherry Raddish

猫山王榴莲布丁 083
Musang King Durian Rice Dumpling

15/72 物候 虹始见

刀鱼饺子 084
Long-Tailed Anchovy Dumpling

桃花泛 086
Sautéed Prawns

谷雨
Grain Rain

16/72 物候 萍始生

大红袍焗豆腐 092
Fried Tofu With Toast Sichuan Pepper

荠菜酱燕窝刀鱼馄饨 095
Long-Tailed Anchovy Wonton With Shepherd's Purse Pesto

17/72 物候 鸣鸠拂其羽

椒麻冲菜牛肉 096
Wagyu Beef With Sichuan Pepper And Pickled Vegetable

丘北辣椒炒羊肚菌 099
Fried Morel With Qiubei Chili

18/72 物候 戴胜降于桑

椒盐刀鱼骨 101
Salt And Pepper Knifefish Bone

焦糖无花果樱桃奶油香草布丁 103
Crème Brulée With Fig Confit And Cherry

立夏 Beginning of Summer

19/72 物候 蝼蝈鸣

奶汤大明湖蒲菜 109
Daminghu Cattail Cream Soup

茉莉花白芦笋汤鸽蛋雪莲 110
White Asparagus Soup With Jasmine

20/72 物候 蚯蚓出

红花汁鳖肚公 112
Braised Cod Fish Maw With Saffron Sauce

豆浆布丁 115
Soy Milk Pudding

21/72 物候 王瓜生

罗勒酱绿竹笋 116
Green Bamboo Shoot With Presto

膏脂高白鲑 119
Slow Baked Coregonus Salmon

小满 Lesser Fullness of Grain

22/72 物候 苦菜秀

陈皮冰淇淋 125
Traditional Orange Peel Sorbet

牡丹黄鱼生 126
Peony Yellow Croaker

23/72 物候 靡草死

夏天的莫奈花园 128
Jardin De Monet

红油猪手 130
Pork Feet With Sichuan Spicy Oil

24/72 物候 麦秋至

天妇罗紫苏鱼子酱 133
Fried Shiso With Caviar

老坛泡菜 134
Asparagus Lettuce Kimchi

芒种 Grain in Beard

25/72 物候 螳螂生

夜香花牛肝菌 140
Porcini With Night Jasmine

渐变慕斯 142
Mousse In Variated Hues

26/72 物候 鵙始鸣

罗马生菜配金钩 145
Grilled Romaine Lettuce With Dried Shrimp

伊比利亚火腿粽子 146
Jamon Iberico Dumpling

27/72 物候 反舌无声

青柠乌鱼蛋冷汤 148
Cuttlefish Roe Cold Soup With Lime

咸蛋黄绿豆饭 150
Green Mongbean Rice With Salted Egg In Lemon Leaves

夏至 Summer Solstice

28/72 物候 鹿角解

普宁豆瓣酱沙姜焗帝王蟹 156
Puning Soy Bean Bake Alaska King Crab

绮霞红菜头沙拉 159
Beetroot Salad

29/72 物候 蜩始鸣

陈皮红豆沙冰棍 161
Frozen Red Bean

百香果慕斯 162
Passionfruit Mousse

30/72 物候 半夏生

蓝调冰粉燕窝 165
Blue Bird Nest

脆皮咸鸡 166
Crispy Salty Chicken

小暑 Lesser Heat

31/72 物候 温风至

一碗胡椒炒软兜 172
Ricefield Eel With Pepper In Stone Pot

芒种虾烧丝瓜 174
Braised Luffa With Dried White Shrimp

32/72 物候 蟋蟀居壁

玫瑰花木姜子金枪鱼腩 177
Litsea Cubeba Oil Tuna Rose

花椒油浸梅童鱼 179
Poach Baby Croaker In Sichuan Pepper Oil

33/72 物候 鹰始鸷

花草焗褐菇 180
Shitake Bouquet

海胆麻酱茄泥 183
Baby Aubergine With Sesame Sauce

大暑 Greater Heat

34/72 物候 腐草为萤

意大利帕尔玛火腿太湖白虾仁炒猫耳朵 188
Orecchiette With White Shrimp And Prosciutto Di Parma

凉瓜河蚌狮子头 190
Freshwater Mussel Meat Ball With Bitter Gourd

35/72 物候 土润溽暑

西瓜泡馍 192
Chill Water Melon Bun

六月黄 194
Steamed Hair Crab

36/72 大雨时行

牛油果酱四川凉面 196
Chill Avocado Noodle With Sichuan Sauce

鲜炖燕窝配牡丹虾 199
Peony Shrimp With Bird Nest

立秋
Beginning of Autumn

37/72 物候 凉风至

虾子茭白 204
Wild Rice Stem With Dried Shrimp Roe

指橙黑叉烧 206
Barbecued Wagyu

38/72 物候 白露降

臭鳜鱼天妇罗配鱼子 209
Tempura Preserved Mandarin Fish

花椒油炒酥藕 211
Quick-fried Lotus With Sichuan Pepper

39/72 物候 寒蝉鸣

拔丝苹果 212
Crispy Toffee Apple

青柠海盐牛肉 214
Beef With Lime Zest And Sea Salt

白露
White Dew

43/72 物候 鸿雁来

老坛泡凤梨 236
Pineapple Pickle

加蚌油鸡枞 238
Geoduck With Fried Yunna Mushroom

44/72 物候 玄鸟归

香辣茴香根 241
Marinated Fennel

松露全素 242
Assort Vegetables With Truffles

45/72 物候 群鸟养羞

姜汁红糖豆腐冻 244
Tofu Pudding With Ginger Syrup

杏汁冰淇淋 247
Almond Ice-Cream

寒露
Cold Dew

49/72 物候 鸿雁来宾

山楂糕和玫瑰鱼子 269
Hawthorn With Rose Caviar

咖喱阿拉斯加蟹 270
Alaska King Crab Curry

50/72 物候 雀入大水为蛤

香茅焗乳鸽 272
Dry Braised Squab With Lemon Grass

松露汁烧鲍鱼配意大利米 275
Risotto With Truffle Oil And Braised Abalone

51/72 物候 菊有黄花

伊比利亚火腿焗百合 277
Bake Lily Bud With Jamon Iberico

大红浙醋烧花胶 278
Braised Fish Maw With Red Vinegar Broth

处暑
The End of Heat

40/72 物候 鹰乃祭鸟

话梅淮山 220
Chinese Yam With Preserved Plum

芫爆蛰头 223
Quick-fried Jelly Fish With Coriander

41/72 物候 天地始肃

桂花糕 224
Osmanthus Cake

红胡椒焗比目鱼 226
Baked Sole With Red Pepper

42/72 物候 禾乃登

红豆沙燕窝 229
Red Bean Soup With Bird Nest

泡菜虾仁 231
Shrimp With Pickle

秋分
Autumnal Equinox

46/72 物候 雷始收声

燕窝南瓜花汤 253
Pumpkin Soup With Bird Nest

萝卜羊肉汆丸子 255
Daikon Soup With Mini Lamb Ball

47/72 物候 蛰虫坯户

赛螃蟹 256
Fish In Crab Styl

山楂覆盆子 259
Hawthorn Raspberry

48/72 物候 水始涸

泉水松茸 260
Stewed Matsutake With Spring Water

枫糖浆渍野樱桃甑糕 262
Steam Glutinous Rice Cake With Maple Syrup Cherry

霜降
Frost's Descent

52/72 物候 豺乃祭兽

青蟹烧龙口粉丝 284
Braised Crab With Mung Bean Vermicelli

帝王蟹焗饭 286
Alaska King Crab Rice

53/72 物候 草木黄落

烧牛肝菌配意大利米面 289
Braised Porcini With Risoni

红花汁栗子白菜 291
Braised Cabbage With Chestnut In Saffron Sauce

54/72 物候 蛰虫咸俯

汕头老鹅肝配鱼子酱 292
Goose Liver With Caviar

霜叶红于二月花 295
Color Of Autumn

立冬

55/72 物候 水始冰

凤尾白菜 300
Crispy Cabbage

墨鱼汁文思豆腐 303
Tofu Julienne In Squid Ink Soup

56/72 物候 地始冻

山楂鹅肝 304
Hawthorn Foie Gras

贝壳邂逅白芝士巧克力 306
White Chocolate Sea Shell

57/72 物候 雉入大水为蜃

蟹糊布丁 309
Crab Pudding

青豆肉饼帝王蟹 310
Steamed Alaska King Crab With Meat Cake

小雪

58/72 物候 虹藏不见

江雪糖醋小排 316
Sweet Pork Ribs At Snowy River

白菜双墩 319
Pickled Cabbage In Dual Flavors

59/72 物候 天气上升

川辣皇转转饼汤 320
Sichuan Spicy Pancake Soup

泡玫瑰萝卜 322
Rose Radish

60/72 物候 闭塞而成冬

青稞石榴沙拉 324
Highland Barley Pomegranate Salad

憋辣菜 327
Pickled Wild Kohlrabi Radish

大雪

61/72 物候 鹖鴠不鸣

黑松露酱烧大响螺黄牛肝菌配意大利米面 332
Braised Conch And Porcini With Truffle And Risoni

摔碎了的蛋 334
Egg

62/72 物候 虎始交

书皮肉饼 337
Crispy Meat Pancake With Cheese

蛋羹秃黄油 338
Steam Hairy Crab Roe With Egg

63/72 物候 荔挺出

男人汤 340
Gentleman's Soup

女人汤 343
Lady's soup

冬至

64/72 物候 蚯蚓结

明炉醋椒鳜鱼 349
Sour and Spicy Mandarin Fish

手磨杏汁炖荷包翅 350
Double Boiled Shark Fin With Almond Cream

65/72 物候 麋角解

焦糖冻柿子 352
Caramelized Iced Persimmon

董氏葱烧海参配葛仙米 354
Dadong Sea Cucumber With Nostoc

66/72 物候 水泉动

红花汁荷包翅佐意大利 30 年香脂醋 357
Braised Shark Fin With Saffron Sauce &
30 Years Balsamic Vinegar

北京酸菜爆浆东莞肉蛋 358
Beijing Sauerkraut With Mini Dired Sausage

小寒

67/72 物候 雁北乡

泡沫乌鱼蛋汤 364
Cuttlefish Eggs Soup

一串串的糖葫芦 367
Hawthorn Skewer

68/72 物候 鹊始巢

合肥炒米泡沫 368
Fried Rice

涮阿拉斯加蟹 370
Alaska King Crab Hot Pot

69/72 物候 雉始雊

赤糖福鼎槟榔芋头 372
Steam Taro With Dark Sugar

葵香鸡蛋萨其马 374
Egg Caramel Treats

大寒

70/72 物候 鸡始乳

干式熟成罗西尼牛排和焦糖肥肝 380
Steak Rossini With Caramel Foie Gras

大董雪糕 382
Da Dong's Ice-Cream

71/72 物候 征鸟厉疾

灌蟹芙蓉蛋 384
L'Oeuf Au Crab

脆柿子沙拉 387
Persimmon Salad

72/72 物候 水泽腹坚

京糕梨丝 389
Hawthorn Cake With Shredded Pears

小吃八款 390
Petit Fours

三九天，北京朔风寒冷，是灰灰的清冷色。立春，一场小雪后，就改了模样。傍晚，南新仓外有『暮色的烟紫』。眼前这抹『一团云烟』的紫色如此浪漫、美好，真想把这抹颜色在琥珀一样的肉皮冻中凝固。

——《**有滋有味的肉皮冻儿**》

During the "Three Nine" days, Beijing's north wind chills to the bone, casting a cool grey hue over the city. With the arrival of the Beginning of Spring, the landscape transforms after a light snowfall. By evening, outside Naxincang (Imperial Granary), there's a "smoky purple" hue in the twilight. The romantic and beautiful shade of purple before your eyes, reminiscent of "a cloud of smoke," is so enchanting that one wishes to capture its essence within an amber-like aspic.

—**The Flavorful Aspic**

立春

Beginning of Spring

THE 1ST PHENOLOGICAL PERIOD
The east wind thaws

The warm eastern wind begins to melt the frozen ground.

01 / 72 物候
东风解冻

东风送暖,大地开始解冻。

种 豆 南 山 下　　草 盛 豆 苗 稀

东晋 · 陶渊明《归园田居》

Traditional Beijing Zhajiang Noodles With Soybean Paste, Pork And Toon

Preparation : 20 mins. Ingredients : cooked hand-pulled noodles 100g, Chinese toon 50g, pork belly 50g, soy bean 50g(cooked), green soy bean 50g(cooked), cucumber 50g (finely sliced), bean sprout 50g(cooked), radish 50g (finely sliced), diced celery 50g (cooked). Seasonings : traditional soybeen paste 50g, soy sauce 30ml, salt 2g, sugar 5g, ground white pepper 1g, shallots 5g, ginger 5g, star anise 5g. Finishing : 3 hrs. Method : 1. Combine soybean paste with soy sauce to dilute. Finely dice pork belly. 2. Heat oil in pan and saute anise till aromatic, add in diced pork belly, chopped ginger and shallots. Add the diluted soybean paste and the other seasonings then simmer for 2 hours. 3. Slightly boil toon then finely chopped. 4. Serve noodles with fried soybean paste, finely chopped toon and the other ingredients.

香椿炸酱面

准备时间:20分钟。食材:手擀面100克,香椿50克,五花肉50克,黄豆50克(煮熟),毛豆50克(煮熟),黄瓜丝50克,豆芽50克(煮熟),水萝卜丝50克,芹菜丁50克(氽水)。调料:干黄酱50克,酱油30毫升,盐2克,白砂糖5克,胡椒粉1克,大料5克,葱、姜各5克。制作时间:3小时。做法:1.干黄酱用酱油澥开,五花肉切成丁。2.起锅放底油,把大料煸出香味,再煸炒五花肉丁和葱、姜,倒入澥好的黄酱和其余调料,小火熬制2小时备用。3.香椿飞水,切末备用。4.将面条煮好,搭配炸酱、香椿和其他面码即可。

Color Combination　色彩搭配:

银烛秋光冷画屏　轻罗小扇扑流萤

唐 · 杜牧《秋夕》

Unfinished Dumpling

Preparation : 5 mins. Ingredients : crayfish 2pcs, Kaluga caviar 15g. Seasonings : wasabi fish roe 5g, soy fish roe 5g. Finishing : 30 mins. Method : 1. Mince crayfish meat to make dumpling wrapper. 2. Form Kaluga caviar into olive shape and place on wrapper to pair with seasonings.

没包完的饺子

准备时间：5分钟。食材：鳌虾2只，卡露伽鱼子15克。调料：芥末鱼子5克，酱油鱼子5克。制作时间：30分钟。做法：1.将鳌虾制成虾胶，做成饺子皮状。2.将卡露伽鱼子做成橄榄状，放在做好的饺子皮中间，搭配调料中的两种鱼子摆盘。

Color Combination　色彩搭配：

THE 2ND PHENOLOGICAL PERIOD
Hibernating insects awaken

Insects that have been
in hibernation start to wake up.

02 / 72 物候
蛰虫始振

— 蛰居的虫类开始苏醒。

天 街 小 雨 润 如 酥　　草 色 遥 看 近 却 无

唐 · 韩愈《早春呈水部张十八员外》

Broad Bean With Preserved Olive Cabbage
Preparation : 5 mins. Ingredients : broad bean 100g, preserved olive cabbage 30g, sweet pea 20g. Seasonings : seasoned soy sauce 2ml, sugar 1g. Finishing : 15 mins. Method : 1.Blanch the broad bean and sweet pea in boiling water, then refreshing in ice water, drain. 2. Finely chopped the preserved olive cabbage then mix with seasonings. 3. Mix the bean with preserved olive cabbage, garnish with sweet pea, serve.

油浸橄榄菜拌蚕豆
准备时间：5 分钟。食材：蚕豆 100 克，橄榄菜 30 克，甜豆 20 克 。调料：鲜酱油 2 毫升，白砂糖 1 克。制作时间：15 分钟。做法：1. 蚕豆和甜豆汆水煮熟后，放入冰水中过凉，备用。2. 将橄榄菜切碎，与调料混合。3. 将蚕豆沥干水分，与步骤 2 拌匀，用甜豆点缀，装盘即可。

Color Combination　色彩搭配：

一颗樱桃天付与　　数声水调人飘逸

宋 · 洪适《满江红 · 席上答叶宪》

Pomegranate Foie Gras

Preparation : 10 mins. Ingredients : foie gras pate 60g. Seasonings : salt 3g, pomegranate juice 100ml, agar agar 1g. Finishing : 1 hr. Method : 1. Heat pomegranate juice and agar agar in a small pot, chill in a flat sheet pan, then move into refrigerator to solidify into jelly sheet. 2. Dice foie gras, season with salt, then wrap in pomegranate jelly sheet.

樱桃花鹅肝

准备时间：10分钟。食材：鹅肝酱60克。调料：盐3克，石榴汁100毫升，琼脂1克。制作时间：1小时。做法：1. 将石榴汁和琼脂放入小锅加热，然后倒入平盘，放入冰箱凝固成石榴啫喱皮。2. 将鹅肝酱切成方块，用盐调味后包裹上石榴啫喱皮即可。

Color Combination 色彩搭配：

THE 3RD PHENOLOGICAL PERIOD
Fish swim beneath the ice

Though the ice hasn't fully melted, fish begin to swim upstream.

03 / 72 物候
鱼陟负冰

虽然冰还没有完全融化,但鱼渐上游。

一条藤径绿　　万点雪峰晴

唐 · 李白《冬日归旧山》

Tofu Spring Onion
Preparation : 5 mins. Ingredients : hard tofu 300g, spring onion 40g. Seasonings : salt 4g, sesame oil 20ml. Finishing : 10 mins. Method : 1.Crush the hard tofu into fine piece, chop spring onion. 2.Mix well the salt, sesame oil with tofu and fine chopped spring onion. 3.Use foil to make tofu ball, ready to serve.

小葱拌豆腐
准备时间:5 分钟。食材:北豆腐 300 克,香葱(也称小葱)40 克。调料:盐 4 克,香油 20 毫升。制作时间:10 分钟。做法:1. 将香葱切末,同时把北豆腐碾碎备用。2. 把盐、香油与碎豆腐混合,放入切好的葱末。3. 拌匀后用保鲜膜做成圆球即可上盘。

Color Combination　色彩搭配:

西塞山前白鹭飞　桃花流水鳜鱼肥

唐 · 张志和《渔歌子》

Fish Fillets In Fermented Rice Wine

Preparation : 10 mins. Ingredients : mandarin fish meat 100g, egg white 1pc. Seasonings : fermented rice wine 50ml, salt 1g, sugar 2g, some corn starch. Finishing : 30 mins. Method : 1. Slice fish to thin fillets and rinse with water. 2. Add salt and sugar, then coat with egg white and corn starch. Blanch fillets in boiling water, let cool and set aside. 3. Boil the rice wine, add starch water to thicken, and then combine with fillets.

香糟春塘片

准备时间：10 分钟 。食材：鳜鱼肉 100 克，鸡蛋 1 个（取蛋清）。调料：糟酒 50 毫升，盐 1 克，白砂糖 2 克，玉米淀粉少许 。制作时间：30 分钟。做法：1. 鳜鱼肉片成薄片，冲水。2. 将冲好水的鱼片加盐、白砂糖打上劲，加入蛋清和玉米淀粉上浆，飞水后冷却备用。3. 糟酒烧开，用玉米淀粉勾芡，再放入鱼片裹匀即可。

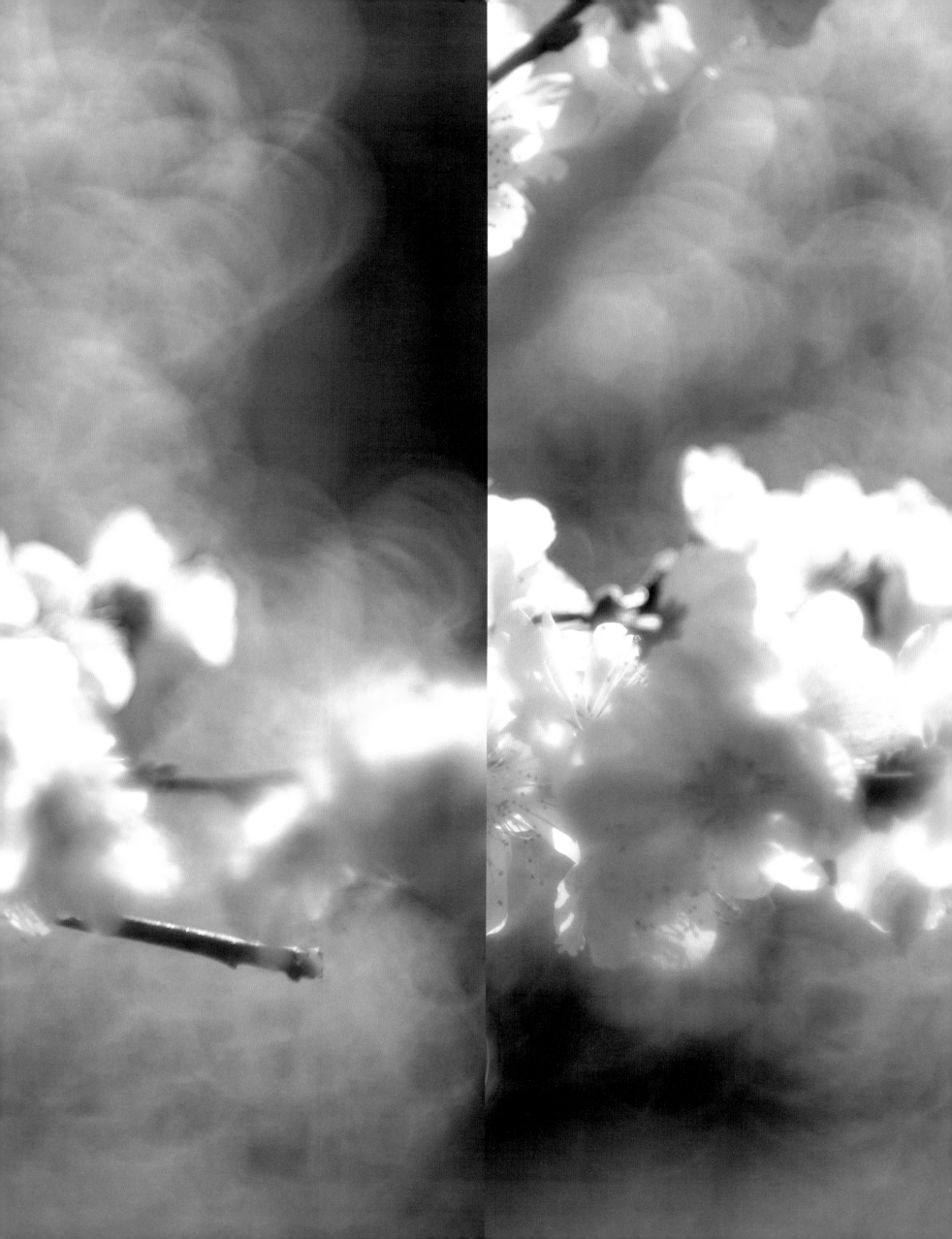

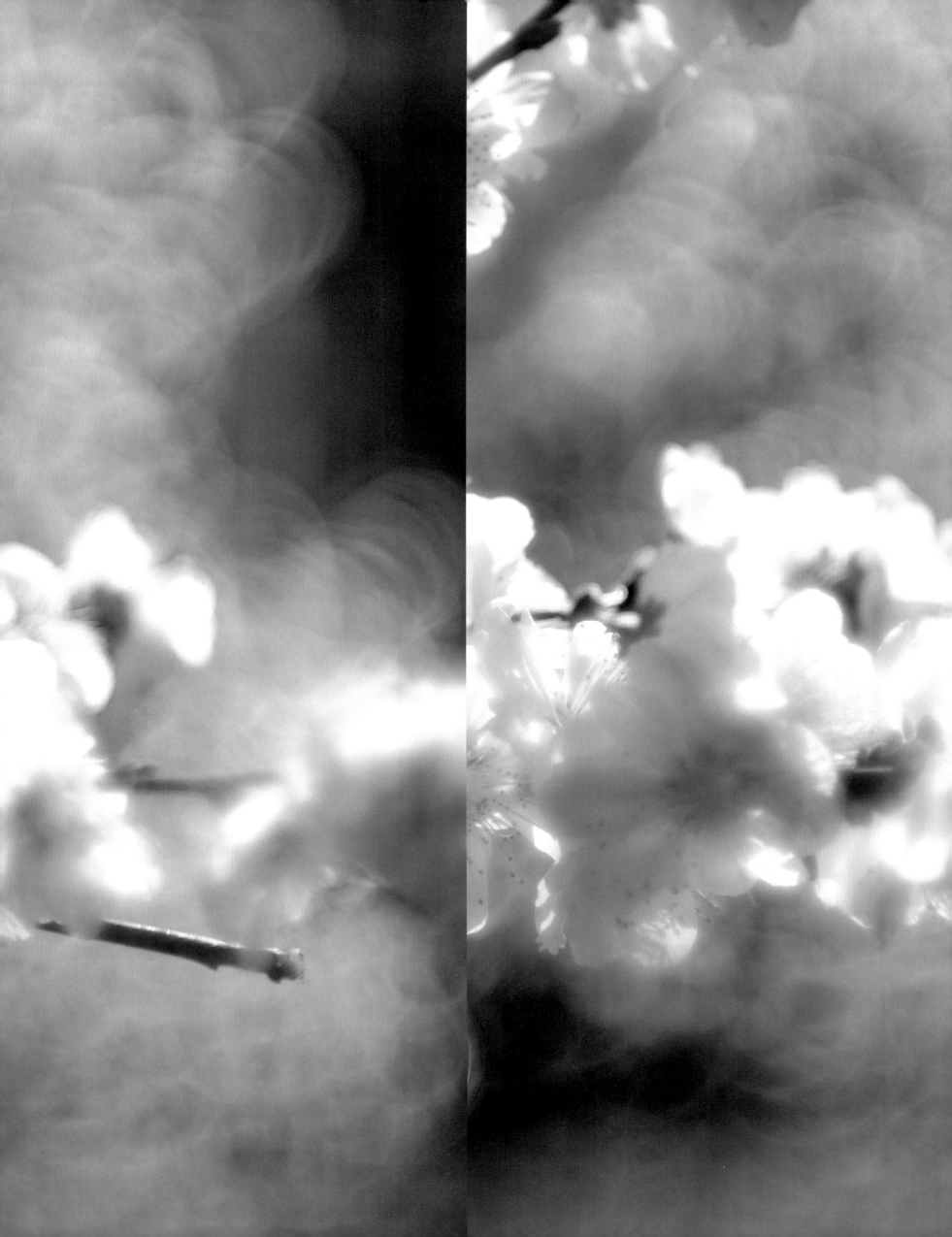

到了『雨水』，江南荠菜最先见青。荠菜刚出嫩芽，最是新春的第一绿。

——《『雨水』日，食春味第一鲜》

By the time of Rain Water, the shepherd's purse in the southern regions is the first to turn green. The tender buds of the shepherd's purse signify the first green of the new spring.

—**On 'Rain Water' Day, Savor the First Freshness of Spring**

雨水

Rain

Water

THE 4TH PHENOLOGICAL PERIOD
Otters worship fish

As the water warms and fish swim upstream, otters begin hunting fish, offering them as a ceremonial tribute before eating.

04 / 72 物候
獭祭鱼

水暖鱼上游,水獭开始捕鱼,先祭而后食。

似烧非因火　如花不待春

唐 · 白居易《和杜录事题红叶》

Spicy Boiled Sole With Dry Aged Sirloin

Preparation : 30 mins. Ingredients : sole fillet 450g, thin sliced dry aged sirloin carpaccio 35g, canola flower 10g. Seasonings : salt water 1000ml, spicy oil 1000ml, ground chili 4g, ground Sichuan pepper 3g, Maggi spicy sauce 10ml, starch 30g, egg white 1pc. Finishing : 30 mins. Method : 1. Soak sole fillet in salt water for 1 min, drain. 2. Mix sole fillet with ground chili, ground Sichuan pepper, Maggi spicy sauce, egg white and starch. 3. Heat spicy oil to 90°C , add sole fillet poach until cooked. Drain. 4. Arrange sole on plate, top with sliced dry aged carpaccio, garnish with canola flower.

无油水煮比目鱼配干式熟成牛肉

准备时间:30分钟。食材:比目鱼450克(取肉),熟成西冷牛肉薄片35克,油菜花10克。调料:盐水1000毫升,水煮油1000毫升,辣椒面4克,花椒面3克,麻辣鲜露10毫升,淀粉30克,蛋清1个。制作时间:30分钟。做法:1.比目鱼肉切块,放入盐水中浸泡1分钟,沥干备用。2.将比目鱼肉用花椒面、辣椒面、麻辣鲜露、淀粉、蛋清拌匀。3.锅中加热水煮油至90°C,放入比目鱼肉烫熟,沥干后装盘。4.将熟成西冷牛肉片码放在鱼肉上,用油菜花点缀即可。

Spicy oil / Ingredients : sunflower seed oil 500ml, canola flower seed oil 200ml, lard 100g, leek 10g, sliced ginger 10g, garlic 10g. Spices : Bhut Jolokia 30g, Sichuan pepper 10g, Angelica Root 3g, black cardamom 3g, clove 3g, cinnamon 3g, star anis 3g, luo han guo 3g, white pepper 3g. Method : 1. Fry leek, sliced ginger, garlic in mixed oil until golden brown. 2. Take away leek, ginger and garlic, add all the spices into oil, turn off the fire, cover for 6 hours, then filter.

水煮油制作 / 食材:葵花籽油500毫升,菜籽油200毫升,猪油100克,大葱段10克,姜片10克,大蒜10克。香料:魔鬼辣椒30克,花椒10克,白芷3克,草果3克,丁香3克,桂皮3克,八角3克,罗汉果3克,白胡椒3克。做法:在锅中将葵花籽油、菜籽油、猪油混合,将大葱段、姜片、大蒜放入锅中以中火炸至金黄色捞出,再放入香料,关火焖6小时,分隔出油分即可。

Color Combination　色彩搭配:

春归翠陌
平莎茸嫩
垂杨金浅

宋·陈亮《水龙吟·春恨》

Sweet Pea Soup With Black Sea Salt
Preparation : 15mins. Ingredients : sweet pea 500g (blanched), chicken soup 350ml. Seasonings : white roux 20g, cream 50ml, salt 5g, sugar 3g, black sea salt 2g. Finishing : 20mins. Method : 1. Blend 400g sweet pea with chicken stock, filter. 2. Season 1 with salt and sugar, heat in low fire, add roux and cream, then mix well. 3. Pour the soup into bowl, garnish with remaining sweet pea and black sea salt.

黑海盐奶油豆蓉汤
准备时间：15分钟。食材：小嫩豆500克（汆水），鸡汤350毫升。调料：黄油炒面20克，淡奶油50毫升，盐5克，白砂糖3克，黑海盐2克。制作时间：20分钟。做法：1. 把400克小嫩豆与鸡汤放入搅拌机打碎，过滤后备用。2. 小火加热豆汤，用盐与白砂糖调味后，加入黄油炒面和淡奶油搅拌。3. 将汤装入碗中，点缀上剩余的小嫩豆，撒上黑海盐即可。

Color Combination　色彩搭配：

THE 5TH PHENOLOGICAL PERIOD
Wild geese fly north

Sensing the signs of spring,
wild geese return from the south.

05 / 72 物候
候雁北

— 大雁感知春信,从南方飞回北方。

绿杨烟外晓寒轻　　红杏枝头春意闹

宋·宋祁《玉楼春》

Crispy Mushroom In Tomato

Preparation : 10 mins. Ingredients : baby Tomato 2 pc, cream cheese 10g, eryngii Mushroom 15g. Seasonings : mayonnaise 15g. Champagne Jelly : Champagne 375ml, gelatin 100g, sugar 125g. Finishing : 30 mins. Method : 1. Mix the Champagne jelly's ingredients together, boil in a low heat, then keep aside. 2. Cut the bottom part of the baby tomato, peel out the seeds by a coffee spoon. 3. Dip the baby tomato into Champagne jelly mixture at least 3 times, keep the tomato in a cool place, until it sets. 4. Peel the eryngii mushroom in thin slices, deep fry until golden colour and crispy, drain. 5. Mix the crispy mushroom with mayonnaise, stuff into baby tomato. 6. Move the stuffed tomato on plate, garnish with cream cheese.

番茄脆菇沙拉

准备时间:10分钟。食材:小番茄2颗,奶油芝士10克,杏鲍菇15克。调料:蛋黄酱15克。香槟酒冻:香槟375毫升,鱼胶粉100克,白砂糖125克。制作时间:30分钟。做法:1.先将制作香槟酒冻的食材混合,煮开后摊凉备用。2.小番茄底部切开一个小口,然后用小勺将番茄心掏净。3.把步骤2的食材浸入步骤1的食材,然后取出,反复3次后,直至番茄外皮挂上一层酒冻,备用。4.杏鲍菇刨成薄片,炸至金黄色后沥干油分。5.把步骤4的食材与蛋黄酱拌匀。6.将拌匀的杏鲍菇填入番茄后上盘,撒上奶油芝士。

Color Combination　色彩搭配:

年来绿树村边月　　夜半清溪梦里身

明 · 陈嘉谋《缺题》

Crispy Celery With Vinaigrette

Preparation : 10 mins. Ingredients : sliced celery 120g. Seasonings : balsamic vinegar 60ml, light soy sauce 2ml, sugar 50g, agar agar 1g, extra virgin olive oil 5ml. Finishing : 1 hr. Method : 1. Heat balsamic vinegar, light soy sauce, sugar and agar agar in a pot, then keep cool wait for it sets. 2. Toss celery with extra virgin olive oil, arrange on plate. 3. Slice vinegar jelly in thin, top on the celery.

油醋汁无渣芹菜

准备时间：10分钟。食材：芹菜丝120克。调料：意大利陈醋60毫升，生抽2毫升，白砂糖50克，琼脂1克，初榨橄榄油5毫升。制作时间：1小时。做法：1.将意大利陈醋、生抽、白砂糖、琼脂混合后加热融化，然后放入盘中冷却，凝固后即醋冻，切丝备用。2.将芹菜丝与初榨橄榄油拌匀上盘，再放上醋冻即可。

Color Combination　色彩搭配：

THE 6TH PHENOLOGICAL PERIOD
Plants start to sprout

Grasses grow and birds sing, marking a good time for plowing.

06 / 72 物候
草木萌动

草长莺飞，可耕之候。

娉娉袅袅十三余　　豆蔻梢头二月初

唐 · 杜牧《赠别二首》

Tender Peas

Preparation : 30 mins. Ingredients : green peas 50g, baby green peas 10g. Seasonings : salt 1g, olive oil 5ml. Finishing : 30 mins. Method : 1. Cook green peas for 20 minutes then place in blender to purée with salt and olive oil. 2. Blanch baby green peas in hot water then refresh in cold water. 3. Mold purée into balls and coat each with baby green peas, plate as shown.

豌豆小时候

准备时间：30 分钟。食材：豌豆 50 克，小嫩豆 10 克。调料：盐 1 克，橄榄油 5 毫升。制作时间：30 分钟。做法：1. 豌豆煮 20 分钟，加入盐、橄榄油，搅打成豌豆泥。2. 小嫩豆焯水后过凉。3. 将豌豆泥做成球状，在外面点缀一层小嫩豆即可。

Color Combination　色彩搭配：

春晖映处花侵户　冬笋成时竹上檐

明 · 唐顺之《和张尚书甬川新修池亭奉母登憩作》

Braised Bamboo Shoot With Fermented Glutinous Rice
Preparation : 5 mins. Ingredients : bamboo shoot 100g. Seasonings : fermented glutinous rice 400ml, sweet osmanthus 3g, salt 5g. Finishing : 10 mins. Method : 1. Cut bamboo shoot in chunk, and then blanch in boiling water. 2. Braise the bamboo shoot together with seasonings, until the seasonings thickening, and ready to serve.

糟煨冬笋
准备时间：5 分钟。食材：冬笋 100 克。调料：糟酒 400 毫升，糖桂花 3 克，盐 5 克。制作时间：10 分钟。做法：1. 将冬笋切滚刀块，氽水备用。2. 将冬笋块与调料一起下锅，以小火烧至酱汁收浓，装盘即可。

Color Combination　色彩搭配：

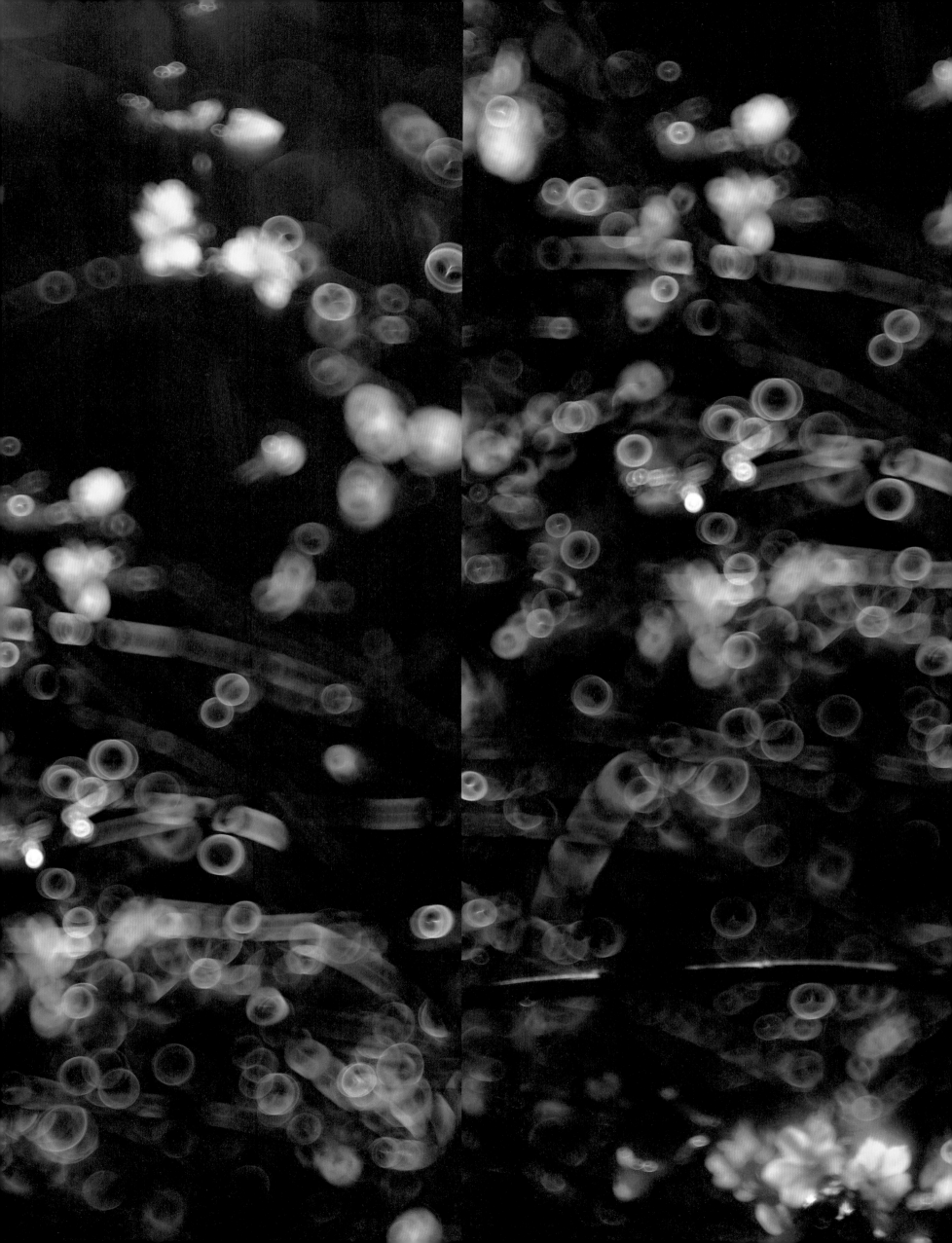

没有吃不了的苦,也没有享不了的福。人生苦短。只是要知道,苦不是人生的全部,苦尽甘来,人生才会丰富多彩,余韵悠扬。

——《有多少苦味可以品尝》

There's no bitterness one can't endure or blessings one can't enjoy. Life is short. One must recognize that suffering isn't the entirety of life. Only through experiencing hardship can one truly appreciate joy, leading to a colorful and fulfilling life.

—*How Many Bitter Flavors Can One Taste*

惊蛰

Insects

Awaken

THE 7TH
PHENOLOGICAL PERIOD
Peach trees blossom

Peach blossoms start to bloom,
signaling the onset of spring.

07 / 72 物候
桃始华

■ 桃花开始生发，自此渐盛。

Fried Shark Fin With Rice Crust
Preparation : 20 mins. Ingredients : shark fin 300g (soaked), bamboo shoot 100g (finely sliced), rice crust 100g, ginger 20g (finely chopped), leek 20g (finely chopped). Seasonings : salt 5g, sugar3g, chicken fat 10ml, chicken stock 200ml. Finishing : 20 mins. Method : 1. Cook shark fin and bamboo shoot with chicken stock, until water evaporate. 2. Heat ginger and leek with chicken fat, add shark fin and bamboo shoot, stir fry. 3. Season with salt and sugar, then arrange on plate. 4. Dress rice crust and serve.

■

春笋锅巴炒翅
准备时间：20 分钟。食材：鱼翅 300 克（泡发），春笋丝 100 克，锅巴 100 克，姜米 20 克，大葱 20 克。调料：盐 5 克，白砂糖 3 克，鸡油 10 毫升，鸡汤 200 毫升。制作时间：20 分钟。做法：1. 鱼翅、春笋丝分别用鸡汤煮入味，备用。2. 锅中放入鸡油，下入大葱、姜米爆香，再放入春笋丝和鱼翅，炒制出"锅气"后，用盐和白砂糖调味，装盘。3. 最后撒入锅巴即可。

燕飞莺语
依约提篮去

宋·韩淲《点绛唇·王园》

Color Combination 色彩搭配：

草长莺飞二月天　　拂堤杨柳醉春烟

清 · 高鼎《村居》

Marinated Mushroon With Caviar

Preparation : 5 mins. Ingredients : afafa 200g, baiyugu(white shimeji) 250g, caviar 75g, quail's egg 1pc (boiled). Seasonings : onion 20g (finely chopped), Maggi 2ml, salt 1g, olive 5ml. Finishing : 30 mins. Method : 1. Cut the top part of the baiyugu, after blanch in boiling water, shock in ice water. Drain. 2. Mix the baiyugu with the seasonings, keep aside for 20 mins. 3. Arrange the afafa and baiyugu on plate, garnish with the quail's egg and caviar.

春草黑鱼子蘑菇沙拉

准备时间：5 分钟。食材：苜蓿芽 200 克，白玉菇 250 克，鱼子酱 75 克，鹌鹑蛋 1 个（煮熟）。调料：洋葱末 20 克，美极鲜 2 毫升，盐 1 克，橄榄油 5 毫升。制作时间：30 分钟。做法：1. 先将白玉菇头用小刀切下，汆水后浸入冰水镇凉，沥干水分备用。2. 把白玉菇头与调料拌匀后腌制 20 分钟入味，然后与苜蓿芽、鹌鹑蛋一起装盘。3. 将鱼子酱撒在鹌鹑蛋表面即可。

THE 8TH
PHENOLOGICAL PERIOD
The oriole sings

The oriole, sensing the warmth of spring, chirps in search of a mate.

08 / 72 物候
仓庚鸣

■ 黄鹂最早感春阳之气,嘤鸣求友。

Sliced Canadian Geoduck Clam With Sichuan Pepper Sprout
Preparation : 5 mins. Ingredients : geoduck 50g, fresh Sichuan pepper sprout 10g. Seasonings : seasoning soy sauce 20ml, soy sauce 2ml. Finishing : 20 mins. Method : 1. Slice geoduck, then quickly blanch in boiling water, sock in ice water. Drain. 2. Mix geoduck with seasoning soy sauce and soy sauce . 3. Quickly blanch fresh Sichuan pepper sprout, dress on geoduck then ready to serve.

■

鲜花椒芽炝加蚌
准备时间:5分钟。食材:加蚌50克,鲜花椒芽10克。调料:蒸鱼豉油20毫升,鲜酱油2毫升。制作时间:20分钟。
做法:1.加蚌片成片,放入沸水中并迅速捞起,再放入冰水中浸泡。2.将加蚌片控干水分,与蒸鱼豉油、鲜酱油拌匀。
3.将步骤2上盘,把快速氽水后的鲜花椒芽点缀在其上即可。

鼎铼也应知此味
莫教姜桂独成功
宋·刘子翚《花椒》

Color Combination 色彩搭配:

此时黄鱼最称美 风味绝胜长桥鲈

明·汪琬《有客言黄鱼事纪之》

Green Garlic With Yellow Croaker

Preparation : 20 mins. Ingredients : yellow croaker 1pc (fillet), green garlic 400g (diced), shallot 200g (sliced), ginger 100g (sliced). Seasonings : Puning soy bean paste 25g, rice spirit 5ml, salt water 1000ml, peanut oil 10ml. Finishing : 20 mins. Method : 1. Soak the yellow croaker in salt water for 1 min, drain and dry. 2. Slice yellow croaker in pieces, mix with Puning soy bean paste. 3. Heat peanut oil in clay pot, fry ginger and shallot, then add 300g green garlic. 4. Arrange yellow croaker on then green garlic, cover and cook for 5 mins. 5. Dress remaining green garlic and rice spirit to serve.

青蒜焗黄鱼

准备时间：20 分钟。食材：大黄鱼 1 条（取肉），青蒜 400 克（切粒），小红葱头 200 克（切片），姜片 100 克。调料：普宁豆瓣酱 25 克，米酒 5 毫升，盐水 1000 毫升，花生油 10 毫升。制作时间：20 分钟。做法：1. 大黄鱼用盐水浸泡 1 分钟后沥干、切块，用普宁豆瓣酱拌匀备用。2. 砂锅中放入花生油，放入姜片、小红葱头片煸炒，放入 300 克青蒜粒，再码放黄鱼，盖盖后焗 5 分钟。3. 撒入剩余的青蒜粒，烹入米酒即可。

Color Combination 色彩搭配：

THE 9TH PHENOLOGICAL PERIOD
Hawks turn into doves

At this time, hawks transform into doves,
and by autumn, doves turn back into hawks.

09/ 72 物候
鹰化为鸠

■ 此时鹰化为鸠，至秋则鸠复化为鹰。

百亩庭中半是苔　桃花净尽菜花开

唐 · 刘禹锡《再过游玄都观》

Canola Flower Salad
Preparation : 10 mins. Ingredients : lamb lettuce10g, balsamic caviar 6g, Parmigiano-Reggiano 3g (grated), canola flower 20g. Seasonings : extra virgin olive oil 3ml, wafer paper 3pcs. Finishing : 10 mins. Method : 1. Toss lamb lettuce with extra virgin olive oil, and wrap together with balsamic caviar, Parmigiano-Reggiano into wafer paper. 2. Seal wafer paper, arrange on plate together with canola flower.

■
油菜花沙拉
准备时间：10 分钟。食材：鸡毛菜 10 克，意大利黑醋鱼子 6 克，帕马森奶酪碎 3 克，油菜花 20 克。调料：特级初榨橄榄油 3 毫升，糯米纸 3 张。制作时间：10 分钟。做法：1. 将鸡毛菜与特级初榨橄榄油拌匀，将帕马森奶酪碎及意大利黑醋鱼子用糯米纸包裹。2. 将糯米纸封口，与油菜花一起装盘。

Color Combination　色彩搭配：

暖雨晴风初破冻　柳眼梅腮　已觉春心动

宋·李清照《蝶恋花》

Fried Baby Octopus With Chives
Preparation : 10 mins. Ingredients : baby octopus 250g (blanched), chives 100g, spring onion 20g (finely chopped), ginger 20g (finely chopped). Seasonings : oyster sauce 3ml, XO sauce 5g, peanut oil 10ml. Finishing : 15 mins. Method : 1. Panfry baby octopus until golden color. 2. Fry ginger and spring onion, XO sauce until fragrant, add baby octopus and chives and oyster sauce, then stir fry with high heat, serve.

韭菜苔炒望潮
准备时间：10 分钟。食材：小望潮 250 克（余水），韭菜苔 100 克，葱米 20 克，姜米 20 克。调料：蚝油 3 毫升，XO 酱 5 克，花生油 10 毫升。制作时间：15 分钟。做法：1. 将小望潮煎至微焦面，备用。2. 锅中煸炒姜米、葱米，倒入 XO 酱，放入小望潮、蚝油及韭菜苔翻炒即可。

Color Combination　色彩搭配：

春分『吃小』，也似乎有一番大周章。个中滋味，如同人情冷暖，冬去春回。人的佳良，与食的佳酿，究竟是怎样的关系呢？

——《春分『吃小』》

Savoring "Small Dishes" during the Spring Equinox seems to follow a grand scheme. The intricate flavors, akin to the warmth and coldness of human emotions, oscillate between the end of winter and the onset of spring. What is the relationship between human nature and the delicacies of food?

—— Savoring Small Dishes During the Spring Equinox

春分

Spring

Equinox

059

THE 10TH PHENOLOGICAL PERIOD
Swallows return

Swallows migrate back from the south.

10 / 72 物候
玄鸟至

▪ 燕子从南方归来。

天分付　使人间草木　尽有春香

宋 · 程必《沁园春 · 寿王运使》

Wagyu With Canola Flower
Preparation : 20 mins. Ingredients : wagyu beef 40g (2mm slices), canola flower 8pcs. Seasonings : steam fish soy sauce 10ml, sea salt 5g, dark soy sauce 2ml, extra virgin olive oil 2ml, water starch 2ml. Finishing : 15 mins. Method : 1. Heat steam fish soy sauce, then thicken by water starch, season with dark soy sauce, keep aside. 2. Toss canola flower with sea salt and extra virgin olive oil. 3. Grill wagyu slices with a blow torch, then roll up the canola flower. 4. Place the wagyu roll on plate, then dress with fish soy sauce.

▪

火灼烟熏雪花牛肉配油菜花
准备时间：20 分钟。食材：雪花牛肉 40 克（2 毫米薄片），油菜花 8 枝。调料：蒸鱼豉油 10 毫升，海盐 5 克，老抽 2 毫升，特级初榨橄榄油 2 毫升，水淀粉 2 毫升。制作时间：15 分钟。做法：1. 蒸鱼豉油加热，用水淀粉勾制薄芡，加入少许老抽调成牛肉汁备用。2. 油菜花用海盐调味，加入橄榄油拌匀，备用。3. 用喷枪将雪花牛肉片炙熟，撒海盐调味后与油菜花一起卷起。4. 将油菜花牛肉卷上盘，配牛肉汁即可。

Color Combination　色彩搭配：

渐觉东风料峭寒　青蒿黄韭试春盘

宋 · 苏轼《送范德孺》

Sweet And Sour Crown Daisy
Preparation : 5 mins. Ingredients : crown daisy 100g, canola flower 5g. Seasonings : sweet and sour vinaigrette 80ml. Finishing : 3 mins. Method : 1. Wash crown daisy and canola flower, drain. 2. Arrange crown daisy and canola flower on plate, dress sweet and sour vinaigrette to serve.

Sweet And Sour Vinaigrette / Ingredients : rice vinegar 500ml, sugar 450g, salt 8g, extra virgin olive oil 10ml. Method: Mix all Ingredients into vinaigrette.

油醋汁童子菜
准备时间：5 分钟。食材：茼蒿苗 100 克，油菜花 5 克。调料：糖醋汁 80 毫升。制作时间：3 分钟。做法：1. 洗净茼蒿苗与油菜花，控干水分。2. 将茼蒿苗及油菜花上盘，浇上糖醋汁即可。

糖醋汁 / 龙门米醋 500 毫升，白砂糖 450 克，盐 8 克，特级初榨橄榄油 10 毫升。做法：将食材混合即可。

Color Combination　色彩搭配：

THE 11TH PHENOLOGICAL PERIOD
Thunder starts to rumble

Thunder, representing the sound of Yang, can be heard.

11/ 72 物候
雷乃发声

■ 雷者阳之声，听见雷鸣。

芳沼徒游比目鱼　　幽径还生拔心草

唐 · 骆宾王《艳情代郭氏答卢照邻》

Steamed Yellow Croaker Roll In Fermented Glutinous Rice Fragrance
Preparation : 5 mins. Ingredients : yellow croaker 1pc (1200g). Seasonings : fermented glutinous rice juice 500ml, sweet osmanthus 3g, salt 5g. Finishing : 40 mins. Method : 1. Scaling yellow croaker, wash and filleted. 2.Mix seasonings together, heat. 3.Slice yellow croaker in long piece, roll and steam for 5-8mins. 4.Dress the sauce and place yellow croaker when ready to serve.

老糟蒸岱衢族大黄鱼卷
准备时间：5 分钟。食材：黄鱼 1 条（约 1200 克）。调料：糟酒 500 毫升，糖桂花 3 克，盐 5 克。制作时间：40 分钟。做法：1. 黄鱼去鳞去内脏，起肉后洗净备用。同时将调料混合成糟汁备用。2．把步骤 1 用布擦干水分后切条，卷成卷，蒸 5~8 分钟。3．糟汁加热，淋在黄鱼卷上即可。

Color Combination　色彩搭配：

寻 花 携 李　　红 漾 轻 舟 汀 柳 外

宋 · 朱敦儒《咸字木兰花 · 寻花携李》

Red Tomato Soup With Scallop

Preparation : 15 mins. Ingredients : scallop 16g (sliced), tomato heart 20g, tomato 380g (diced), cabbage 280 (diced), potato 100g (diced), carrot 100g (diced), celery 120 (diced), onion 150g (diced), rose caviar 30g. Seasonings : salt 6g, sugar 50g, roux 20g, tomato paste 230g, water 2000ml. Finishing : 90 mins. Method : 1. Fry tomato, cabbage, potato, carrot, celery and onion in a soup pot. Add tomato paste and water, then simmering for 1hour, pass sieve. 2. Heat tomato soup with roux to thicken, season with salt and sugar, keep warm. 3. Place scallop, tomato heart and rose caviar in bowl plate, pour the tomato soup to serve.

红漾番茄带子汤

准备时间：15 分钟。食材：带子 16 克（薄片），番茄心 20 克，番茄丁 380 克，圆白菜丁 280 克，土豆丁 100 克，胡萝卜丁 100 克，芹菜丁 120 克，洋葱丁 150 克，玫瑰鱼子 30 克。调料：盐 6 克，白砂糖 50 克，黄油炒面 20 克，番茄酱 230 克，水 2000 毫升。制作时间：90 分钟。做法：1. 将番茄丁、圆白菜丁、土豆丁、胡萝卜丁、芹菜丁及洋葱丁在锅中炒香，再加入番茄酱，煸炒 5 分钟后加入水，熬制 1 小时后过筛备用。2. 把番茄汤与黄油炒面加热，以盐和白砂糖调味，保温。3. 盘中放入带子片、番茄心与玫瑰鱼子，倒入番茄汤即可。

Color Combination 色彩搭配：

THE 12TH PHENOLOGICAL PERIOD
Lightning appears

Lightning, representing the light of Yang, begins to flash as the Yang energy increases.

12 / 72 物候
始电

■ 电者阳之光,阳气微则光不见,阳气渐升,有了闪电。

雨 洗 娟 娟 嫩 叶 光　　风 吹 细 细 绿 筠 香

宋 · 苏轼《少年游》

Dragon Whiskers Noodles With Potherb Mustard And Razor Clam
Preparation : 15 mins. Ingredients : dragon whiskers noodles 50g, razor clam 80g, chicken stock 600ml, potherb mustard 20g. Seasonings : salt 3g, ground white pepper 3g, peanut oil 8ml. Finishing : 20 mins. Method : 1. Blanch razor clam and peel the meat. 2. Heat peanut oil, stir fry potherb mustard, add chicken stock, season with salt and ground white pepper, pour in bowl. 3. Boil dragon whiskers noodles for 1min, drain and put into the bowl with razor clam.

■

雪菜蛏子龙须贡面
准备时间:15 分钟。食材:龙须贡面 50 克,蛏子 80 克,鸡汤 600 毫升,雪菜 20 克。调料:盐 3 克,白胡椒粉 3 克,花生油 8 毫升。制作时间:20 分钟。做法:1. 先将蛏子氽水取肉备用。2. 锅中加热花生油,放入雪菜煸炒,加入鸡汤后用盐及白胡椒粉调味,倒入碗中。3. 龙须贡面用沸水煮 1 分钟沥干水分,放入碗中,最后放上蛏子肉即可。

Color Combination　色彩搭配:

几处早莺争暖树
谁家新燕啄春泥

唐·白居易《钱塘湖春行》

Fried Morel And Broad Bean With Shepherd's Purse Pesto
Preparation : 5 mins. Ingredients : morel 2pcs (blanched), broad bean 15g (blanched). Seasonings : butter 5g, shepherd's purse pesto 5g, salt 3g. Finishing : 10 mins. Method : 1. Panfry morel with butter, stir-fry broad bean, season with salt and keep warm. 2. Place shepherd's purse pesto on plate, then arrange morel and broad bean.

荠菜酱蚕豆羊肚菌
准备时间：5分钟。食材：羊肚菌2个（氽水），蚕豆15克（氽水）。调料：黄油5克，荠菜酱5克，盐3克。制作时间：10分钟。做法：1. 羊肚菌用黄油两面煎熟，蚕豆炒熟，以盐调味备用。2. 盘上放上荠菜酱，再分别放上羊肚菌及蚕豆即可。

Shepherd's Purse Pesto / Ingredients : shepherd's purse 500g (blanched), pinenut 100g (toasted), extra virgin olive oil 200ml, salt 3g. Method: Blend all Ingredients until smooth, season with salt.

荠菜酱 / 食材：荠菜500克（氽水），松仁100克（烘烤），初榨橄榄油200毫升，盐3克。做法：将所有食材放入搅拌机打碎，用盐调味。

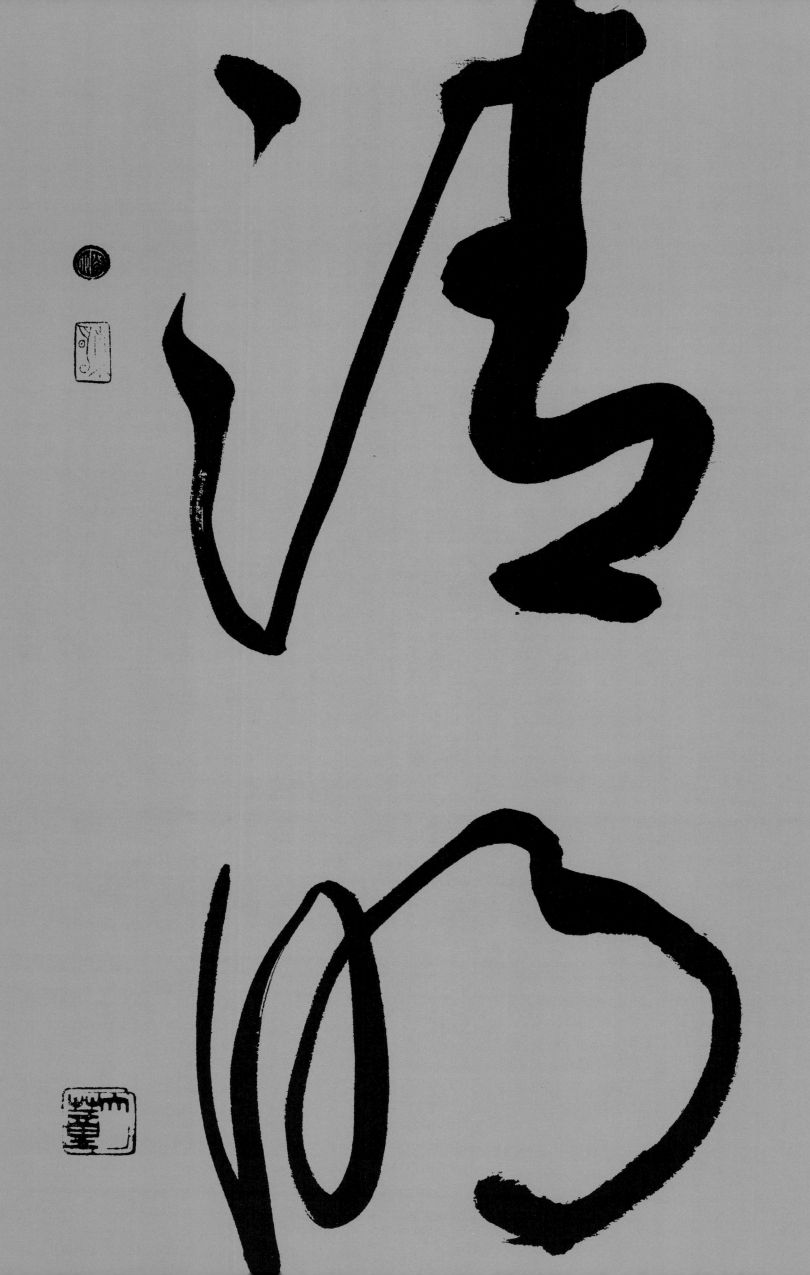

如有一夜春风，杨树花就开了。樱桃小萝卜跟着杨树花，也就这几天的吃鲜。过后，只是留下小萝卜红唇皓齿的样子，期待明年。

——《春天的杨花小萝卜》

After a night of spring winds, poplar flowers bloom. Cherry radishes, following the poplar flowers, enjoy their brief season of freshness. Soon, only memories of the red-lipped, pearly-toothed radishes remain, awaiting their return the next year.

—Springtime Poplar Flower and Cherry Radish

清明

Pure

Brightness

THE 13TH PHENOLOGICAL PERIOD
Paulownia trees blossom

Paulownia trees begin to flower.

13 / 72 物候
桐始华

桐树开花。

桃花浅深处　似匀深浅妆

唐 · 元稹《桃花》

Braised Pork Belly And Green Bean Starch Sheet
Preparation : 30 mins. Ingredients : green bean starch sheet 180g (soaked), bamboo shoot 60g (sliced), pork belly 80g (sliced), peach blossom 3g, crispy beetroot sheet 1pc. Seasonings : soy sauce 10ml, chicken stock 300ml, sugar 5g, star anis 3pcs. Finishing : 30 mins. Method : 1. Fry star anis with pork belly, add soy sauce, sugar, chicken stock and bamboo shoot. 2. Add green bean starch sheet and braise until absorb the stock, then place into bowl plate, cover with crispy beetroot sheet and garnish with peach blossom.

Crispy Beetroot Sheet / Ingredients: flour 10g, starch 18g, sunflower seed oil 15ml, beetroot juice 1ml, water 60ml. Method : Mix all the ingredients, fry until water evaporate.

五花肉烧粉皮桃花泛
准备时间：30 分钟。食材：粉皮 180 克（泡软），笋片 60 克，五花肉片 80 克，桃花 3 克，红菜头薄脆一片。调料：酱油 10 毫升，鸡汤 300 毫升，白砂糖 5 克，大料 3 颗。制作时间：30 分钟。做法：1. 将大料与五花肉片煸香，放入酱油、白砂糖、鸡汤及笋片，最后放入粉皮。2. 粉皮烧至软烂，装入碗中，盖上红菜头薄脆，点缀上桃花即可。

红菜头薄脆 / 食材：面粉 10 克，淀粉 18 克，葵花籽油 15 毫升，紫菜头汁 1 毫升，水 60 毫升。做法：把食材混合后放入不粘锅中煎至水分蒸发即可。

Color Combination　色彩搭配：

波纹碧皱　曲水清明后

宋 · 晏几道《清平乐 · 波纹碧皱》

Steam Bamboo Shoot And Shepherd's Purse Bun

Preparation : 30 mins. Ingredients : bamboo shoot 600g (blanched), shepherd's purse 200g (blanched), bun dough 60g. Seasonings : salt 2g, fried shallot oil 60ml. Finishing : 30 mins. Method : 1. Diced bamboo shoot and finely chopped shepherd's purse, mix with salt and fried shallot oil. 2. Roll bun dough into 1mm thin, fill with the bamboo shoot mixture, then steam for 8mins to serve.

Bun Dough / Ingredients : dumping flour 300g, water 150ml. Method : Mix dumping flour and water, knead the dough until hard, set to use.

问政山笋荠菜包子

准备时间：30 分钟。食材：问政山笋 600 克（汆水），荠菜 200 克（汆水），包子面皮 60 克。调料：盐 2 克，葱油 60 毫升。制作时间：30 分钟。做法：1. 问政山笋切丁备用，荠菜切末，并与盐、葱油混合成馅备用。2. 将包子面皮擀成 1 毫米薄，填入馅料后隔水蒸 8 分钟即可。

包子面皮 / 食材：饺子面 300 克，清水 150 毫升。做法：和面并将其揉成偏硬一点的面团即可。

Color Combination　色彩搭配：

THE 14TH PHENOLOGICAL PERIOD
Field mice turn into quails

14 / 72 物候
田鼠化为鴽

Shade-loving field mice retreat underground.

喜阴的田鼠,回到了地下。

红绿修岸容
冰雪净空界

宋·叶适《丁氏东屿书房》

Shredded Cherry Raddish

Preparation : 5 mins. Ingredients : baby radish 30g Seasonings : sugar 20g, salt 0.5g, rice vinegar 15ml. Finishing : 15 mins. Method : 1. Clean and shred baby radish. 2. Marinate baby radish with rice vinegar, salt, and sugar. Plate as shown.

樱桃萝卜

准备时间:5 分钟。食材:樱桃萝卜 30 克。调料:白砂糖 20 克,盐 0.5 克,米醋 15 毫升。制作时间:15 分钟。做法:1. 将洗净后的樱桃萝卜切蓑衣花刀备用。2. 将改好刀的樱桃萝卜加米醋、盐、白砂糖拌匀,入味即可。

Color Combination 色彩搭配:

081

秋露白如玉　　团团下庭绿

唐 · 李白《 古风其二十三》

Musang King Durian Rice Dumpling
Preparation : 30 mins. Ingredients : Musang king durian 400g (peeled). Seasonings : glutinous rice flour 250g, wheat starch 70g, lard 35g, sugar 50g, water 190ml, matcha powder 2g. Finishing : 20 mins. Method : 1. Mix all seasonings then knead into dough, portion in 12g each. 2. Musang king durian portion in 10g each, then wrap into the matcha dough, form into ball. 3. Steam for 8 mins, let cool and serve.

猫山王榴莲布丁
准备时间：30 分钟。食材：猫山王榴莲肉 400 克。调料：糯米粉 250 克，熟澄面 70 克，猪油 35 克，白砂糖 50 克，水 190 毫升，抹茶粉 2 克。制作时间：20 分钟。做法：1. 将调料混合，揉成面团后分成数个各 12 克的小面团备用。2. 将猫山王榴莲肉分成各 10 克的小份，用小面团包裹后隔水蒸 8 分钟，摊凉即可。

Color Combination　色彩搭配：

THE 15TH
PHENOLOGICAL PERIOD
Rainbows appear

With thin clouds and the sun shining through the rain, rainbows can be seen.

15 / 72 物候
虹始见

■ 云薄漏日，日穿雨影，可见彩虹。

Long-Tailed Anchovy Dumpling
Preparation : 40 mins. Ingredients : long-tailed anchovy 250g (minced), shrimp meat 50g, cuttlefish 250g (minced), dumpling wrapper 500g . Seasonings : lard 50g, sesame oil 20 ml, salt 10g, ground white pepper 2g, leek and giner water 180ml. Finishing : 20 mins. Method : 1. Mix long-tailed anchovy, shrimp meat and cuttlefish with seasonings into stuffing. 2. Fill the stuffing into dumpling wrapper, fold into dumpling. 3. Cook the dumpling into boiling water for 8 mins, drain and serve.

■

刀鱼饺子
准备时间：40 分钟。食材：刀鱼肉 250 克，虾仁 50 克，墨鱼肉馅 250 克，饺子面皮 500 克。调料：猪油 50 克，芝麻油 20 毫升，盐 10 克，白胡椒粉 5 克，葱姜水 180 毫升。制作时间：20 分钟。做法：1. 将刀鱼肉、虾仁、墨鱼肉馅与调料混合成馅料。2. 将馅料用饺子面皮包成饺子，放入沸水煮 8 分钟，沥干即可上盘。

清水飘芙蓉
元宝落玉盘
现代·胡秉言《饺子》

Color Combination 色彩搭配：

桃之夭夭　灼灼其华

先秦·佚名《桃夭》

Sautéed Prawns

Preparation : 10 mins. Ingredients : king prawn 2 pcs (around 400g), ginger julienne 30g . Seasonings : salt 2g, sugar 25g, vegetable oil 30ml, pure water 100ml. Finishing : 25 mins. Method : 1. Clean king prawn, peel off the tharm on the back and cut off the spear on the head. 2. Stir-fry ginger julienne in a pan, add king prawn then fry both side into golden colour. 3. Pour the pure water into 2, then season with salt and sugar. 4. Turn to low heat, reduce the sauce to syrup consistence. Ready to serve.

桃花泛

准备时间：10分钟。食材：对虾 2只（约 400克），姜丝 30克。调料：盐 2克，白砂糖 25克，色拉油 30毫升，纯净水 100毫升。制作时间：25分钟。做法：1．对虾去掉虾枪和虾线后洗净备用。2．把姜丝用色拉油煸出香味，放入对虾煎至金黄色。3．加入纯净水，以盐和白砂糖调味。4．调成小火，待汁收浓即可。

Color Combination　色彩搭配：

菱蒿满地，芦芽芽短。
枸杞头嫩，草头清香，马兰花开，荠菜出尖。
还有刀剪的柳芽，串串榆钱。白芹白，苋菜红。

——《春吃芽、夏吃瓜、秋吃果、冬吃根》

The ground is covered in abundant mugwort, with short sprouts of cattail. The goji berries are tender, and the grass emits a refreshing aroma. Nasturtium blossoms in full bloom, while shepherd's purse shoots emerge gracefully. Additionally, there are willow shoots meticulously trimmed, and clusters of elm seeds. The celery appears pure white, and the amaranth displays a vibrant red.

— **In Spring, We Indulge In Tender Shoots; In Summer, We Savor Succulent Melons; In Autumn, We Relish Bountiful Fruits; In Winter, We Enjoy Nourishing Roots**

谷雨

Grain

Rain

THE 16TH PHENOLOGICAL PERIOD
Duckweed begins to grow

Duckweed starts to float and propagate on water.

16 / 72 物候
萍始生

浮萍随水漂浮而生。

三千两钟乳　八百斛胡椒

宋 · 刘克庄《杂咏一百首 · 其三十八 元载》

Fried Batter / Ingredients : flour 300g , water 310ml, oil 320ml . Method : Mix well all ingredients.

Fried Tofu With Toast Sichuan Pepper
Preparation : 30 mins. Ingredients : lactone tofu 300g, fresh Sichuan pepper sprout 20g (blanched). Seasonings : fried batter 800g, brine 1l, flour 200g, red Sichuan pepper 250g, Sichuan pepper oil 3ml, salt 2g. Finishing : 40 mins. Method : 1. Slice lactone tofu into 1.5cmX5cm pieces, then soak into brine for 20mins, drain. 2. Dry lactone tofu, toss with flour and dip into fried batter, deep fry in 210℃ until golden color. 3. Toast red Sichuan pepper, pour into plate, arrange with fried tofu, then heat the plate on fire for 3mins. 4. Toss fresh Sichuan pepper sprout with Sichuan pepper oil and salt, garnish on the top of tofu.

大红袍焗豆腐
准备时间：30 分钟。食材：内酯豆腐 300 克，花椒芽 20 克（余水）。调料：酥炸糊 800 克，盐水 1 升，面粉 200 克，大红袍花椒 250 克，花椒油 3 毫升，盐 2 克。制作时间：40 分钟。做法：1. 将内酯豆腐切成长 5 厘米、宽 1.5 厘米的四方条，浸入饱和盐水中泡 20 分钟备用。2. 让豆腐吸干水分，蘸干面粉后挂酥炸糊，以 210℃ 热油炸至金黄色，沥干。3. 大红袍花椒在锅中焗香，倒入盘中，放上炸豆腐，然后放在炉灶上以小火焗 3 分钟。4. 花椒芽用花椒油和盐拌一下，点缀在豆腐上即可。

酥炸糊 / 食材：面粉 300 克，水 310 毫升，色拉油 320 毫升。做法：将食材混合成糊即可。

Color Combination　色彩搭配：

烟 销 日 出 不 见 人　　欸 乃 一 声 山 水 绿

唐 · 柳宗元《渔翁》

Shepherd's Purse Pesto / Ingredients : shepherd's purse 500g (blanched), pine nut 100g (toasted), extra virgin olive oil 200ml, salt 3g. Method : Blend all Ingredients until smooth, season with salt.

Long-Tailed Anchovy Wonton With Shepherd's Purse Pesto
Preparation : 30 mins. Ingredients : long-tailed anchovy 100g (filleted), bird nest 30g (soaked), wonton wrapper 20g. Seasonings : lard 30g, chicken stock 100ml, shepherd's purse pesto 20g. Finishing : 30 mins. Method : 1. Fine chopped long-tailed anchovy fillet, mix with lard. 2. Steam bird nest for 10mins. 3. Fill the long-tailed anchovy mixture into wonton wrapper, fold into wonton, cook in boiling water for 5 mins, drain. 4. Arrange the wonton, bird nest and shepherd's purse pesto on plate, pour in hot chicken stock.

荠菜酱燕窝刀鱼馄饨
准备时间：30 分钟。食材：刀鱼 100 克（起肉），燕窝 30 克（泡发），馄饨皮 20 克。调料：猪油 30 克，鸡汤 100 毫升，荠菜酱 20 克。制作时间：30 分钟。做法：1. 将刀鱼肉剁碎，与猪油调合成馅备用。2. 燕窝蒸 10 分钟备用。3. 刀鱼馅用馄饨皮包好，放入沸水中煮 5 分钟，沥干水分。4. 盘中放入刀鱼馄饨、燕窝和荠菜酱，再淋上热鸡汤即可。

荠菜酱 / 食材：荠菜 500 克（氽水），松仁 100 克（烘烤），初榨橄榄油 200 毫升，盐 3 克。做法：将所有食材放入搅拌机打碎，用盐调味。

Color Combination　色彩搭配：

THE 17TH
PHENOLOGICAL PERIOD
Doves coo and preen

Doves sing and preen themselves, signaling people to sow seeds.

17 / 72 物候
鸣鸠拂其羽

斑鸠鸣叫梳理羽毛，提醒人们播种。

驰隙流年　恍如一瞬星霜换

宋 · 张抡《烛影摇红 · 上元有怀》

Wagyu Beef With Sichuan Pepper And Pickled Vegetable
Preparation : 20 mins. Ingredients : preserved vegetable 50g, wagyu beef 30g. Seasonings : salt 0.5g, Maggi sauce 1ml, mustard sauce 1g, mustard oil 2ml, green Sichuan pepper pesto 10g. Finishing : 20 mins. Method : 1. Wash vegetables, pat dry and finely chop. Season with salt, Maggi sauce, mustard sauce and mustard oil. 2. Panfry beef to medium well with both sides scortched. Serve with vegetables and green Sichuan pepper pesto as shown.

椒麻冲菜牛肉
准备时间：20 分钟。食材：冲菜 50 克，雪花牛肉 30 克。调料：盐 0.5 克，美极鲜酱油 1 毫升，芥末膏 1 克，芥末油 2 毫升，椒麻酱 10 克。制作时间：20 分钟。做法：1. 冲菜洗净焯水，过凉后剁碎，加入所有调料拌匀备用。2. 将雪花牛肉煎至六成熟，有焦面，和椒麻酱一起上盘。

Green Sichuan Pepper Pesto / Ingredients: green Sichuan pepper 50g (no seed), spring onion (only green part, finely chopped). Seasoning: sun flower seed oil 15ml. Method: Put all ingredients and sun flower seed oil into food processor, blend to paste.

椒麻酱 / 食材：青花椒 50 克（去籽），小葱 30 克（绿色部分切碎）。调料：葵花籽油 15 毫升。做法：将所有食材和调料一起放入打碎机中打成酱即可。

Color Combination　色彩搭配：

菌耳遍沃野　琼芝亦芬芳

元 · 吾丘衍《菌耳遍沃野》

Fried Morel With Qiubei Chili
Preparation : 1 hr. Ingredients : morel 200g (blanched), fish maw 100g (soaked), Qiubei chili 100g (soaked). Seasonings : salt 5g, butter 10g. Finishing : 15 mins. Method : 1. Stuff fish maw into morel, pan fry with butter, season with salt. 2. Add Qiubei chili, stir fry in high heat until fragrant, serve to plate.

丘北辣椒炒羊肚菌
准备时间：1 小时。食材：羊肚菌 200 克（汆水），花胶 100 克（泡发），丘北辣椒 100 克（泡水）。调料：盐 5 克，黄油 10 克。制作时间：15 分钟。做法：1. 将花胶酿入羊肚菌，用黄油煎香，用盐调味。2. 加入丘北辣椒，用大火炒香呈焦面即可。

Color Combination　色彩搭配：

THE 18TH PHENOLOGICAL PERIOD
Hoopoes perch on mulberry trees

18 / 72 物候
戴胜降于桑

Hoopoes can be seen resting on mulberry branches.

戴胜鸟栖息在桑树上。

墙燕惊三叠
刀鱼送两旗

宋 · 黄彦平《泊舟兜率寺呈王承可》

Salt And Pepper Knifefish Bone

Preparation : 5 mins. Ingredients : whole knifefish 1 pc. Seasonings : salt and pepper 3g, tempura flour 10g. Finishing : 25 mins. Method : 1. Steam to cook fish, pat dry and separate meat from bone. Keep bone for use. 2. Coat bone with tempura batter and deep fry till golden, season with salt and pepper. Plate as shown.

椒盐刀鱼骨

准备时间：5 分钟。食材：刀鱼 1 条。调料：椒盐 3 克，天妇罗粉 10 克。制作时间：25 分钟。做法：1. 刀鱼蒸熟后去肉，鱼骨备用。2. 在刀鱼骨外蘸一层天妇罗粉，炸至金黄色，用椒盐调味即可。

Color Combination　色彩搭配：

樱桃半点红　怜美景　惜芳容

宋 · 晏几道《阮郎归 · 晚妆长趁景阳钟》

Crème Brulée With Fig Confit And Cherry
Preparation : 30 mins. Ingredients : dried fig 100g, cherry 200g (sliced), egg yolk 2pcs, vanilla 1pc, milk 200ml, cream 200ml. Seasonings : sugar 200g, water 150ml, syrup 500ml. Finishing : 2hrs. Method : 1. Cook dried fig in syrup, let cool. Mix egg yolk with cream. 2. Heat sugar with 50ml water, add 100g water when it caramelizes, mix well and pour into bowl. 3. Heat milk and vanilla in a pot to 90°C , pour into cream mixture, stir well. 4. Pour the mixture into bowl then bake in 100°C for 1hr. 5. Arrange cherry and fig confit on the crème brulée.

焦糖无花果樱桃奶油香草布丁
准备时间：30 分钟。食材：无花果（干）100 克，樱桃 200 克（切肉），蛋黄 2 个，香草夹 1 根，牛奶 200 毫升，淡奶油 200 毫升。调料：白砂糖 200 克，水 150 毫升，糖水 500 毫升。制作时间：2 小时。做法：1. 将无花果放入糖水中煮 30 分钟，摊凉。淡奶油和蛋黄混合，备用。2. 将 200 克白砂糖加入 50 毫升水中烧制成焦糖色，再加入 100 毫升水制作成焦糖浆，倒入碗中备用。3. 锅中加入香草荚和牛奶，加热至 90°C，倒入蛋黄奶油再混合均匀，摊凉备用。4. 将摊凉的香草奶油倒入碗中，放入带水的烤盘里，在预热 100°C 的烤箱中烤 1 小时。5. 将樱桃和无花果放在烤好的布丁上即可。

Color Combination　色彩搭配：

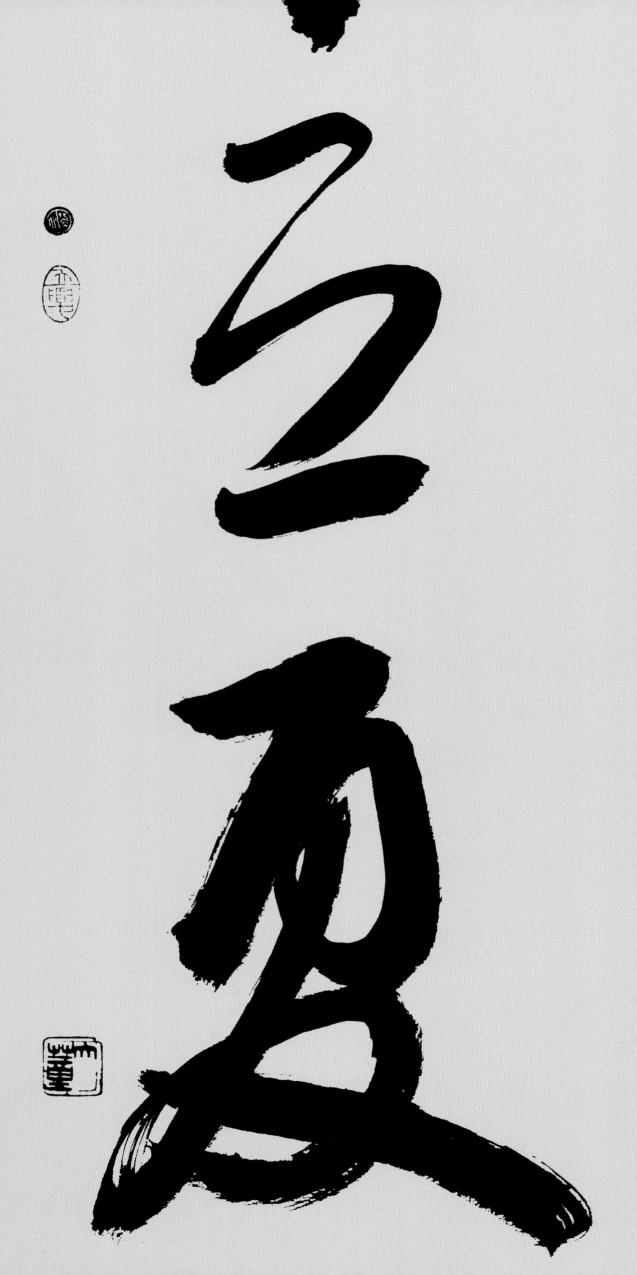

初夏食得一苦笋，今年苦夏必定不苦，定有滋有味。

——《和黄山谷说苦》

At the start of summer, tasting a bitter bamboo shoot ensures that the upcoming harsh summer won't feel as bitter, promising a flavorful season.

—**Discussing Bitterness with Huang Tingjian**

立夏

Beginning of Summer

THE 19TH PHENOLOGICAL PERIOD
Mole Crickets sing

19 / 72 物候

蝼蝈鸣

水晶帘动微风起
满架蔷薇一院香

唐·高骈《山亭夏日》

蝼蝈，蛙也，田间听取蛙声一片。

Mole crickets, believed to make the sound of frogs, fill the fields with their croaking.

Daminghu Cattail Cream Soup
Preparation : 5 mins. Ingredients : cattail 150g. Seasonings : chicken stock 150ml, salt 1g, ground white pepper 0.5g. Finishing: 15 mins. Method : 1. Trim roots and peel stalks of cattail. 2. Blanch with hot water, pat dry, and plate. 3. Season chicken stock with salt and ground pepper then pour over cattails and serve.

奶汤大明湖蒲菜
准备时间：5 分钟。食材：蒲菜 150 克。调料：鸡汤 150 毫升，盐 1 克，白胡椒粉 0.5 克。制作时间：15 分钟。做法：1. 蒲菜去掉外面的皮和根部。2. 将蒲菜焯水，晾凉，打卷，装盘。3. 鸡汤用盐、白胡椒粉调味，浇在蒲菜卷上，完成装盘。

Color Combination　色彩搭配：

芦 笋 锥 犹 短　　凌 澌 玉 渐 融

唐 · 元稹《生春二十首（丁酉岁 · 凡二十章）》

White Asparagus Soup With Jasmine
Preparation : 10 mins. Ingredients : white asparagus 200g (diced), pigeon egg 2pcs (boiled), Chinese honeylocust spine 5g (soaked), jasmine blossom 3g. Seasonings : butter 10g, salt 2g, white roux 10g, chicken stock 300ml. Finishing : 20 mins. Method : 1. Peel the egg white of pigeon egg, steam the Chinese honeylocust spine for 5 mins. 2. Stir fry white asparagus with butter, add chicken stock to boil, then blend until smooth. 3. Add white roux to thicken the soup, season with salt, add pigeon egg white and Chinese honeylocust spine. 4. Pour the soup into bowl, garnish with jasmine blossom.

茉莉花白芦笋汤鸽蛋雪莲
准备时间：10 分钟。食材：白芦笋 200 克（切粒），鸽子蛋 2 颗（煮熟），雪莲子 5 克（泡发），茉莉花 3 克。调料：黄油 10 克，盐 2 克，黄油炒面 10 克，鸡汤 300 毫升。制作时间：20 分钟。做法：1. 鸽子蛋取蛋清，雪莲子蒸 5 分钟备用。2. 白芦笋用黄油煎香，加入鸡汤，烧开后用打碎机打碎，备用。3. 汤里加入黄油炒面后以盐调味，和鸽子蛋清、雪莲子一起倒入碗中。4. 食用时撒上茉莉花即可。

THE 20TH PHENOLOGICAL PERIOD
Earthworms emerge

Earthworms come out, sensing the warmth.

20 / 72 物候
蚯蚓出

蚯蚓感阳气而出。

西真宴罢群仙醉
千尺黄云错紫霞

宋·陈造《鹧鸪天》

Braised Cod Fish Maw With Saffron Sauce
Preparation : 10 mins. Ingredients : fish maw 60g. Seasonings : oil 500ml, superior thick consommé 100ml, saffron 1g, salt 2g, sugar 3g, starch 50g. Finishing : 3hrs 30 mins. Method : 1. Soak the dry fish maw in 100°C hot oil for 1 hr, drain. 2. Rub the fish maw with starch, clean out remaining oil, then soak in 80°C hot water for 2 hours, soak in ice water. 3. Drain the fish maw, then slice in strip, heat together with superior thick consommé. 4. Add saffron and remaining seasonings, then serve.

红花汁鳖肚公
准备时间：10分钟。食材：鳖鱼肚60克。调料：食用油500毫升，浓汤100毫升，藏红花1克，盐2克，白砂糖3克，生粉50克。制作时间：3小时30分钟。做法：1. 将鳖鱼肚用100°C的食用油浸泡1小时，将油分沥干。2. 把鳖鱼肚抹上生粉搓洗，将剩余油分吸走，用80°C的水浸泡2小时，再泡入冰水中，备用。3. 把步骤2切成长条，放入烧热的浓汤中。4. 放入藏红花，以盐、白砂糖调味即可。

For Superior Thick Consommé, please see "Braised Cabbage with Chestnut in Saffron Sauce". (P291)

备注：浓汤制作请参考红花汁栗子白菜。（P291）

Color Combination 色彩搭配：

年年今夜 月华如练

宋 · 范仲淹《御街行 · 秋日怀旧》

Soy Milk Pudding
Preparation : 30 mins. Ingredients : soy milk 150ml, spring onion 10g (finely chopped). Seasonings : gelatine sheet 3pcs (soaked), salt 5g. Finishing : 2 hrs. Method : 1. Boil soy milk, add gelatine sheet and salt, stir well, then pipe in a balloon, keep in refrigerator for set. 2. Place spring onion in plate, then top with soy milk pudding.

豆浆布丁
准备时间：30 分钟。食材：豆浆 150 毫升，小葱 10 克（切细）。调料：鱼胶片 3 片（泡软），盐 5 克。制作时间：2 小时。做法：1. 将豆浆煮沸，加入泡好的鱼胶片和盐，搅拌均匀，灌入气球后放入冰箱冷藏定型。2. 小葱切碎铺在盘底，将冷藏的半球形豆腐放在葱碎上即可。

THE 21ST PHENOLOGICAL PERIOD
Bottle gourds grow

Bottle gourds (from the gourd family) start to climb and grow.

21 / 72 物候
王瓜生

新绿小池塘
风帘动
碎影舞斜阳

宋·周邦彦《风流子·新绿小池塘》

王瓜（葫芦科）的蔓藤开始攀爬生长。

Green Bamboo Shoot With Presto
Preparation : 10 mins. Ingredients : green bamboo shoot 100g, white balsamic caviar 5g, basil 3g. Seasonings : pesto 50g. Finishing : 10 mins. Method : 1. Slice green bamboo shoot into 2.5cm X 2.5cm cube. Arrange on plate. 2. Dress with pesto, then garnish with white balsamic caviar and basil.

Pesto / Ingredients: basil 50g (finely chopped), pine nut 10g. Seasonings: olive oil 40ml, salt 2g. Method: Add basil, pine nut and seasonings into food processor and blend until smooth.

罗勒酱绿竹笋
准备时间：10 分钟。食材：绿竹笋 100 克，意大利白波沙米克爆珠 5 克，罗勒叶嫩尖 3 克。调料：罗勒酱 50 克。制作时间：10 分钟。做法：1. 先将绿竹笋切成长 2.5 厘米、宽 2.5 厘米的方丁，上盘。2. 淋上罗勒酱，码放上罗勒叶嫩尖及意大利白波沙米克爆珠。

罗勒酱 / 食材：罗勒叶 50 克（切碎），松仁 10 克。调料：橄榄油 40 毫升，盐 2 克。做法：将罗勒叶、松仁与调料放入搅拌机打成酱即可。

Color Combination 色彩搭配：

最爱芦花经雨后 一蓬烟火饭鱼船

宋·林逋《咏秋江》

Slow Baked Coregonus Salmon

Preparation : 20 mins. Ingredients : coregonus salmon 1pc around 700g (scaled and filleted). Seasonings : Puning soy paste 50g (blended), wasabi 30g, salt 50g, Chinese sprit 30ml, ice water 2.5l. Finishing : 20 mins. Method : 1. Mix Salt and Chinese sprit into ice water, then soak the coregonus salmon for 5mins. 2. Dry the coregonus salmon and bake in oven (80℃) for 10mins, then slice in rectangle pieces. 3. Arrange the Puning soy paste on plate, then place the coregonus salmon pieces, garnish with the wasabi on top.

膏脂高白鲑

准备时间：20 分钟。食材：高白鲑 1 条约 700 克（去骨、去鳞、内脏）。调料：普宁豆瓣酱 50 克（打碎），青芥末 30 克，盐 50 克，白酒 30 毫升，冰水 2.5 升。制作时间：20 分钟。做法：1. 先将冰水、盐、白酒混合，将高白鲑放入浸泡 5 分钟。2. 将浸泡好的高白鲑抹干水分，放入预热 80℃的烤箱中烤制 10 分钟后取出，切块。3. 将普宁豆瓣酱挤在盘上，放上高白鲑，最后挤上青芥末即可。

Color Combination　色彩搭配：

石榴花开,夏至夏半,老百姓的吃食倒变得简单了。拍个黄瓜,麻酱茄泥,过水凉面,浓荫下扇着蒲扇,自在惬意。

——《天棚鱼缸石榴树,老爷肥狗胖丫头》

When pomegranate flowers bloom, around midsummer, people's diets become simpler. Smashing a cucumber, eggplant puree with sesame paste, or cold noodles - enjoying these under the dense shade while fanning oneself feels utterly comfortable and content.

— **High-Ceilinged Residence With Fish Tanks And Pomegranate Trees, The Master, Chubby Dogs, And Plump Maidens**

小满

Lesser

Fullness

of

Grain

THE 22ND PHENOLOGICAL PERIOD
Bitter herbs bloom

22 / 72 物候
苦菜秀

Bitter herbs flower, acquiring their taste from the fire element.

苦菜开花，感火气而生苦味。

锦瑟无端五十弦
一弦一柱思华年

唐·李商隐《锦瑟》

Traditional Orange Peel Sorbet
Preparation : 20 mins. Ingredients : dried orange peels 30g, preserved orange peels 20g. Seasonings : egg 1pc, sugar 60g, whipped cream 100ml, milk 180ml, milk powder 2g, corn starch 3g, gelatin 1pc. Finishing : 13 hrs. Method : 1. Soak dried orange peels in water until soft and dice. 2. Combine 1 and all seasonings until smooth then freeze for 12 hours. 3. Process frozen mixture into fine sorbet then form it with spoon to olive shape, garnish with shredded orange peels.

陈皮冰淇淋
准备时间：20 分钟 。食材：陈皮 30 克，蜜制橙皮 20 克。调料：鸡蛋 1 个，白砂糖 60 克，淡奶油 100 毫升，牛奶 180 毫升，奶粉 2 克，玉米淀粉 3 克，鱼胶片 1 片。制作时间：13 小时。做法：1. 陈皮用水泡软，切成粒备用。2. 将陈皮粒与所有调料混合均匀后装入冰淇淋桶，速冻 12 小时。3. 冻好的冰淇淋用冰霜机打一遍，做成橄榄状，搭配蜜制橙皮装盘。

Color Combination 色彩搭配：

唯有牡丹真国色　花开时节动京城

唐 · 刘禹锡《赏牡丹》

Peony Yellow Croaker
Preparation : 30 mins. Ingredients : yellow croaker 1pc (filleted). Marinade: salt 100g, sugar 400g, mineral water 50ml, orange juice 200ml, beetroot juice 80ml. Seasonings : Puning soy bean paste 25g, garlic oil 5ml. Finishing : 100 mins. Method : 1. Blend Puning soy bean paste with garlic oil, pour into bowl. 2. Mix the marinade, then soak the yellow croaker fillet, keep in refrigerator for 90 mins. 3. Drain and dry the fillet, slice in butterfly shape, arrange in plate. 4. Serve with garlic soy bean paste.

牡丹黄鱼生
准备时间：30 分钟。食材：大黄鱼 1 条（去骨取肉）。腌料：盐 100 克，白砂糖 400 克，矿泉水 50 毫升，橙汁 200 毫升，红菜头汁 80 毫升。调料：普宁豆瓣酱 25 克，蒜油 5 毫升。制作时间：100 分钟。做法：1. 普宁豆瓣酱与蒜油搅打成蓉状，放入碗中备用。2. 将腌料混合，将大黄鱼肉浸入腌料中，放入冰箱冷藏 90 分钟。3. 将鱼肉沥干，用刀片成蝴蝶片，在盘中码放成牡丹花形状。4. 食用时搭配普宁蒜酱。

Color Combination　色彩搭配：

THE 23RD PHENOLOGICAL PERIOD
Shade-loving plants wither

Plants that prefer shade wilt under the intense sun.

23 / 72 物候
靡草死

喜阴的植物靡草，经受不了烈日而凋亡。

红紫斗芳菲　　满园张锦机

宋·张抡《菩萨蛮》

Jardin De Monet
Preparation : 20 mins. Ingredients : Botan-ebi 1pc, beetroot jelly 20g, cooked egg yolk 20g, molecular orange caviar 20g. Seasonings : salt 1g. Finishing : 30 mins. Method : 1. Defrost Botan-ebi, season with salt, mash egg yolk. 2. Plate egg yolk as base, top with Botan-ebi, then garnish with remaining ingredients.

夏天的莫奈花园
准备时间：20分钟。食材：牡丹虾1只，红菜头冻20克，熟鸡蛋黄20克，橙汁鱼籽20克。调料：盐1克。制作时间：30分钟。做法：1. 牡丹虾化冻后去外壳，用盐调味。熟鸡蛋黄碾碎备用。2. 用鸡蛋黄铺底，把拌好的虾码放在上面，用红菜头冻和橙汁鱼籽装饰。

Color Combination 色彩搭配：

知否　知否　应是绿肥红瘦

宋 · 李清照《如梦令 · 昨夜雨疏风骤》

Pork Feet With Sichuan Spicy Oil

Preparation : 20 mins. Ingredients : pork feet half pc, asparagus lettuce 50g, peanut 5g (crusted). Seasonings : spicy broth 2l, Sichuan spicy red oil 10ml, salt 2g. Finishing : 3 hrs. Method : 1. Blanch pork feet, then simmer in spicy broth for 2 hours, debone and slice thin. 2. Thin slice asparagus lettuce, boil in water to cook, season with salt. 3. Arrange slice pork feet and asparagus lettuce on plate. Dress Sichuan spicy red oil and peanut to serve.

红油猪手

准备时间：20 分钟。食材：猪蹄半只，莴笋 50 克，花生碎 5 克。调料：麻辣卤水 2 升，红油 10 毫升，盐 2 克。制作时间：3 小时。做法：1. 猪蹄焯水，放入麻辣卤水中卤制 2 小时，剔骨，片成薄片。2. 将莴笋片成薄片，焯水后加盐入味。3. 将猪蹄片和莴笋片装盘，滴上红油，撒上花生碎即可。

Color Combination　色彩搭配：

THE 24TH PHENOLOGICAL PERIOD
Wheat harvest begins

Summer wheat ripens, marking the wheat harvest season.

24 / 72 物候
麦秋至

此时夏麦成熟，故曰麦秋。

炮笋烹鱼饱飧后
拥袍枕臂醉眠时

唐·白居易《初致仕后戏酬留守牛相公并呈分司诸寮友》

Fried Shiso With Caviar

Preparation : 5 mins. Ingredients : shiso leaf 1pc, caviar 3g, shiso blossom 2g. Seasonings : fried batter 100g. Finishing : 5 mins. Method : 1.Dip shiso leaf into fried batter, then fry in 180 °C until golden color, drain. 2. Garnish with caviar and shiso blossom (Check fried batter recipe from pumpkin soup with bird nest, P253).

天妇罗紫苏鱼子酱

准备时间：5 分钟。食材：紫苏叶 1 片，鱼子酱 3 克，紫苏花 2 克。调料：脆炸糊 100 克。制作时间：5 分钟。做法：1. 紫苏叶蘸裹脆炸糊，以 180℃油温炸至金黄，沥干油。2. 将鱼子酱及紫苏花放在其上即可（脆炸糊请参考燕窝南瓜花汤，P253）。

Color Combination 色彩搭配：

松际露微月　清光犹为君

唐 · 常建《宿王昌龄隐居》

Asparagus Lettuce Kimchi

Preparation : 20 mins. Ingredients : asparagus lettuce 300g. Seasonings : salt 2g, rock sugar 10g, Sichuan pepper 3g pickle juice 300ml, Chinese liquor 15ml. Finishing : 3 hrs 30 mins. Method : 1.Cut asparagus lettuce to 1cm cube. 2.Combine salt, rock sugar, Sichuan pepper, and liquor to marinate asparagus lettuce for 3 hours. 3.Remove asparagus lettuce from marination and plate.

老坛泡菜

准备时间：20 分钟。食材：青笋 300 克。调料：盐 2 克，冰糖 10 克，四川花椒 3 克，泡菜汤 300 毫升，白酒 15 毫升。制作时间：3 小时 30 分钟。做法：1. 青笋切 1 厘米方丁备用。2. 泡菜汤加盐、冰糖、四川花椒、白酒调味，将青笋放入泡菜汤中浸泡 3 小时。3. 捞出泡好的青笋，装盘即可。

Color Combination　色彩搭配：

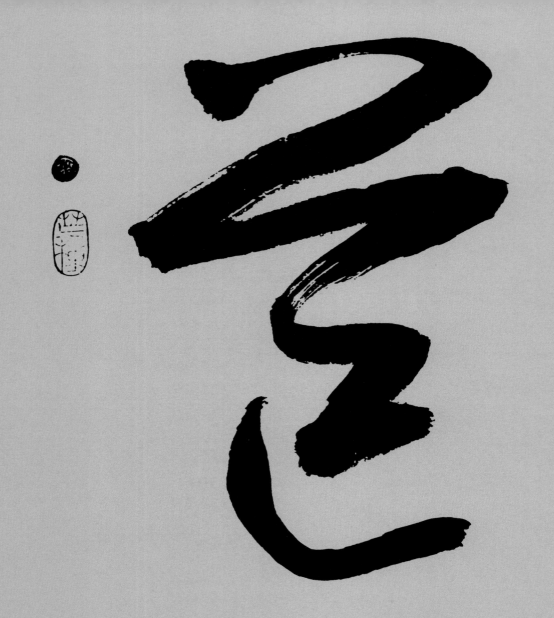

烹饪的滋味不只是食材和调配料的化学物理反应。加入空间这个调料,用时间去烧煮,烹饪的滋味就是人类的从前和今后。

——《成都的『廊桥』和『松云泽』》

The essence of cooking is not merely the chemical and physical reactions between ingredients and seasonings. By introducing space as an ingredient and simmering it with time, the essence of cooking captures the tales of humanity's past and future.

— Chengdu's 'The Bridge' and 'Song Yun Ze'

芒种

Grain in Beard

THE 25TH PHENOLOGICAL PERIOD
Praying mantises hatch

Sensing temperature changes, mantises break out of their shells.

25 / 72 物候
螳螂生

感知气温变化，螳螂破壳而出。

首夏犹清和　　芳草亦未歇

南北朝 · 谢灵运《游赤石进帆海》

Porcini With Night Jasmine
Preparation : 5 mins. Ingredients : porcini 80g, night jasmine 15g. Seasonings : salt 2g. Finishing : 15 mins. Method : 1. Panfry porcini then set aside. 2. Wash and clean night jasmine then saute. 3. Plate fried porcini, season with salt, and garnish with night jasmine.

夜香花牛肝菌
准备时间：5 分钟。食材：牛肝菌 80 克，夜香花 15 克。调料：盐 2 克。制作时间：15 分钟。做法：1. 牛肝菌煎熟备用。2. 夜香花洗净，炒熟。3. 将煎好的牛肝菌撒上盐，用夜香花点缀装盘。

Color Combination　色彩搭配：

梨花菊蕊不相饶　娇黄带轻白

宋 · 赵彦端《好事近 · 朱户闭东风》

Mousse In Variated Hues

Preparation : 30 mins. Ingredients : Kraft Philadelphia cheese 100g. Seasonings : whipped cream 25ml, gelatin 20g, Korean refined sugar 30g, honey 10ml, lemon 1pc, mango nectar 30ml, saffron water 5ml, passion fruit jam 10g, curry powder 3g, orange juice 10ml. Finishing : 40 mins. Method : 1. Mix cheese and all seasonings to create multi-shaded mousse. 2. Cut mousse into cube with variated hues from light to dark.

渐变慕斯

准备时间：30分钟。食材：卡夫菲利芝士100克。调料：淡奶油25毫升，凝胶片20克，韩糖30克，蜂蜜10毫升，黄柠檬1个，芒果浓缩汁30毫升，藏红花水5毫升，百香果酱10克，咖喱粉3克，橙汁10毫升。制作时间：40分钟。做法：1. 将芝士化开，加入调料做成不同颜色、不同口味的慕斯。2. 把做好的慕斯切成方块，颜色由浅到深装盘。

Color Combination　色彩搭配：

THE 26TH PHENOLOGICAL PERIOD
Shrikes start to chirp

26 / 72 物候
鵙始鸣

Shrikes become active on tree branches.

伯劳鸟开始在枝头活动。

绿云翻翠浪
水急转前溪

宋·刘过《临江仙·四景》

Grilled Romaine Lettuce With Dried Shrimp

Preparation : 10 mins. Ingredients : romaine lettuce 200g, dried shrimp 5g, shallot 50g, small lemons 1pc. Seasonings: dried shrimp roe 50g, salt 1g, olive oil 30ml. Finishing : 20 mins. Method : 1. Saute romaine lettuce in olive oil until golden brown. 2. Shallot in halfs and pan-fry to cook. 3.Arrange romaine lettuce and shallot on serving plate, finish with dried shrimp, shrimp roe, salt, garnish by small lemons.

罗马生菜配金钩

准备时间：10 分钟。食材：罗马生菜 200 克，金钩 5 克，小红葱头 50 克，小柠檬 1 个。调料：虾子 50 克，盐 1 克，橄榄油 30 毫升。制作时间：20 分钟。做法：1. 将罗马生菜用橄榄油煎熟，煎至有焦面。2. 小红葱头去皮，从中间切开，煎熟。3. 将罗马生菜与煎好的小红葱头装盘，撒上虾子、盐，点缀上小柠檬（切半）。

Color Combination　色彩搭配：

采丝缠粽动嘉辰　浴殿风生画扇轮

宋·王珪《端午内中帖子词·皇帝阁》

Jamon Iberico Dumpling

Preparation : 10 mins. Ingredients : sliced jamon Iberico 60g, sushi rice 40g, dried pratia grass 5g. Seasonings : sushi vinegar 5ml. Finishing : 5 mins. Method : 1. Mix sushi vinegar with warm sushi rice. 2. Fill the sushi rice into sliced jamon Iberico, then roll into cone shape, tie up with dried pratia grass.

伊比利亚火腿粽子

准备时间：10分钟。食材：伊比利亚火腿片60克，寿司饭40克，马莲草5克。调料：寿司醋5毫升。制作时间：5分钟。做法：1. 将寿司醋拌入热的寿司饭中，摊凉备用。2. 将寿司米饭包入火腿片，将其卷成粽子状并用马莲草捆绑。

Color Combination　色彩搭配：

THE 27TH PHENOLOGICAL PERIOD
The mockingbird stops singing

The mockingbird bird ceases its calls.

27 / 72 物候
反舌无声

■ 反舌鸟停止了鸣叫。

春慵恰似春塘水　　一片縠纹愁

宋 · 范成大《眼儿媚 · 酣酣日脚紫烟浮》

Cuttlefish Roe Cold Soup With Lime
Preparation : 15 mins. Ingredients : cuttlefish roe 25g, celery 30g. Seasonings : lemon juice 2ml, lemon zest 1g, salt 1g, clear broth 150ml. Finishing : 40 mins. Method : 1. Bring clear broth to a boil then add celery, lemon juice, lemon zest, and salt. Pour in blender to blend then sieve and chill. 2. Blanch cuttlefish roe with hot water, let cool then place in chilled broth and serve.

■

青柠乌鱼蛋冷汤
准备时间：15分钟。食材：乌鱼蛋25克，芹菜30克。调料：柠檬汁2毫升，柠檬碎1克，盐1克，清汤150毫升。制作时间：40分钟。做法：1. 清汤烧开，放入芹菜、柠檬汁、柠檬碎，用盐调味， 之后用打碎机打碎，镇凉。2. 将乌鱼蛋焯水后过凉，装入器皿，浇上冷汤后装盘。

Color Combination　色彩搭配：

绿遍山原白满川
子规声里雨如烟

宋·翁卷《乡村四月》

Green Mongbean Rice With Salted Egg In Lemon Leaves
Preparation : 5 mins. Ingredients : green mongbean 200g, rice 200g, salted duck egg yolk 1pc, some lemon leaves. Finishing : 40 mins. Method : 1. Cook rice with mongbeans. 2. Mix cooked beans and rice, form into balls, stuffed with salted duck egg yolk then wraped with lemon leaves.

咸蛋黄绿豆饭
准备时间：5 分钟。食材：绿豆 200 克，大米 200 克，高邮鸭蛋黄 1 个，少许柠檬叶。制作时间：40 分钟。做法：1. 将绿豆、大米蒸成绿豆饭。2. 把蒸好的绿豆饭做成饭团，将鸭蛋黄裹在饭团中，用柠檬叶包好，装盘。

Color Combination 色彩搭配：

没有心，『怜』子就没有情，莲子所有的美好都在芯里，芯的苦，就在经营，经营什么？经营它的甜。
原来苦和甜是对比的。
没了苦，哪里有甜呢？

——《莲心，怜心》

Without the heart, 「compassion」 loses its essence, just like the lotus seed loses its beauty. The bitterness of the lotus plumule serves a purpose - it nurtures the sweetness within. Bitterness and sweetness are contrasting flavors that complement each other. Without bitterness, where would sweetness come from?

— **Lotus Plumule, Compassionate Heart**

夏至

Summer

Solstice

155

THE 28TH PHENOLOGICAL PERIOD
Deer shed their antlers

Male deer, influenced by the Yin energy, begin to shed their antlers.

28 / 72 物候
鹿角解

阳性的鹿角，得阴气而开始脱落。

蟹螯即金液　　糟丘是蓬莱

唐 · 李白《月下独酌》

Puning Soy Bean Bake Alaska King Crab

Preparation : 30 mins. Ingredients : Alaska king crab 1pc, garlic 200g, sliced ginger 200g, leek 200g, green garlic 50g (diced)，galangal 50g (finely chopped), coriander 50g (finely chopped). Seasonings : Puning soy bean paste 50g, peanut oil 10ml, rice spirit 5ml. Finishing : 30 mins. Method : 1. Steam Alaska king crab for 5 mins to medium rare, then peel the crab meat. 2. Mix crab meat with Puning soy bean paste and coriander. Keep in refrigerator for 10 mins. 3. Heat peanut oil in clay pot, fry sliced ginger and leek, add garlic and arrange the crab meat on the top. Cover for 4 mins. 4. Dress galangal and green garlic, dress with rice spirit while it serves.

普宁豆瓣酱沙姜焗帝王蟹

准备时间：30 分钟。食材：阿拉斯加蟹一只，蒜子 200 克，姜片 200 克，大葱段 200 克，青蒜粒 50 克，沙姜末 50 克，香菜末 50 克。调料：普宁豆瓣酱 50 克，花生油 10 毫升，广东米酒 5 毫升。制作时间：30 分钟。做法：1. 阿拉斯加蟹蒸制 5 分钟至五成熟，取出净蟹肉，备用。2. 将蟹肉与普宁豆瓣酱、香菜末拌匀，放入冰箱冷藏 10 分钟。3. 砂锅中放入花生油，放入姜片、大葱段，煸出香味后放入蒜子。再码放上蟹肉，盖上盖焗制 4 分钟。4. 最后撒入沙姜末及青蒜粒，喷上少量广东米酒即可。

Color Combination 色彩搭配：

天外绮霞迷海鹤
日边红树艳仙桃

唐·李绅《寿阳罢郡日，有诗十首》

Beetroot Salad
Preparation : 10 mins. Ingredients : beetroot 30g, blueberry 10g, carrot 10g, shallot 10g. Seasonings : salt 2g, sugar 2g, mayonnaise 3g. Finishing : 30 mins. Method : 1. Cook beetroot, carrot and shallot. 2. Diced beetroot. 3. Mix 2 with all the seasonings, then arrange on plate, dress with blueberry.

绮霞红菜头沙拉
准备时间：10 分钟。食材：红菜头 30 克，蓝莓 10 克，胡萝卜 10 克，红葱 10 克。调料：盐 2 克，白砂糖 2 克，沙拉酱 3 克。制作时间：30 分钟。做法：1. 将红菜头与红葱、胡萝卜一起煮熟。2. 把煮熟的红菜头切成粒。3. 把所有调料拌在一起，装盘，用蓝莓点缀。

Color Combination　色彩搭配：

THE 29TH PHENOLOGICAL PERIOD
Cicadas start to sing

Cicadas, also known as summer insects, buzz loudly.

29 / 72 物候
蜩始鸣

蜩，蝉也，知了鼓翼而鸣。

Frozen Red Bean
Preparation : 10 mins. Ingredients : dried orange peel 100g, red bean 900g. Seasonings : sugar 200g, water 800ml. Finishing : 3 hrs. Method : 1. Simmer 200g red bean together with 100g sugar and 600ml water in a pot for 40mins. 2. Soak the dried orange peel in warm water until soft, then chopped, at the same time, steam the remaining red bean, sugar and water for 2.5hrs. 3. Pass the boiled red bean to a fine sift, to make purée. 4. Mix the chopped dried orange peel, steamed red bean and red bean purée, put into a long mould. 5. Put the mould into deep freezer until it freezes.

陈皮红豆沙冰棍
准备时间：10 分钟。食材：陈皮 100 克，红豆 900 克。调料：白砂糖 200 克，水 800 毫升。制作时间：3 小时。做法：1. 将 200 克红豆与 100 克白砂糖、600 毫升水倒入锅中，以中火煮 40 分钟，沥干水分备用。2. 陈皮泡软切碎，同时将剩余的红豆、白砂糖、水放入碗中蒸 2.5 小时。3. 将步骤 1 打碎，过筛后变成豆沙。4. 将步骤 3 与陈皮和蒸好的红豆混合，装入棒冰模具。5. 把步骤 4 放入冰箱冷冻成棒冰即可。

不为陈皮开膈冷
却留霜子到青春
明·林大钦《园果六首·柑》

Color Combination 色彩搭配：

緋黄紫绿献态度　纤洪浓淡敷条枚

宋・陈淳《和林叔己咏扬守福寿林塘之韵》

Passionfruit Mousse

Preparation : 30 mins. Ingredients : passion fruit 4pcs, white chocolate chips 300g, cream 200ml (whipped), gelatine sheet 2pcs (soaked), milk 100ml, white chocolate shell 8pcs, mango powder 30g. Seasonings : passionfruit juice 200ml, sugar 50g. Finishing : 2hrs. Method : 1. Melt 100g white chocolate chips, mix with mango powder, keep warm.2. Dip white chocolate shell into mango chocolate, which make the shell yellow color. 3. Heat milk with 1 pc of gelatine sheet, then pour into remaining white chocolate chips, stir until melt. 4. Slowly fold the whipped cream into white chocolate, keep in refrigerator. 5. Cut passionfruit and take out the flesh, mix with passionfruit juice, sugar and gelatine. 6. Pipe white chocolate mousse into chocolate shell, then top with passionfruit mixture. 7. keep in refrigerator for 1hr until it set.

百香果慕斯

准备时间：30分钟。食材：百香果4个，白巧克力片300克，淡奶油200毫升（打发），鱼胶片2片（泡软），牛奶100毫升，白巧克力壳8个，芒果粉30克。调料：百香果汁200毫升，白砂糖50克。制作时间：2小时。做法：1. 100克白巧克力片隔水化开，与芒果粉混合后备用。2. 将白巧克力壳浸入芒果巧克力中，使其外壳变成黄色，备用。3. 牛奶烧开放入鱼胶片，待鱼胶片融化后倒入剩余白巧克力中，让白巧克力充分融化。4. 将打发的淡奶油加入白巧克力中慢慢搅拌制成慕斯，冷藏备用。5. 百香果切开，挖出果肉籽，与百香果汁、白砂糖及鱼胶片搅拌均匀。6. 将白巧克力慕斯挤入白巧克力壳中冷却定型，用勺子挖出多余的巧克力再注入百香果酱。7. 把百香果慕斯放入冰箱冷藏1小时定型即可。

Color Combination 色彩搭配：

THE 30TH PHENOLOGICAL PERIOD
Half-summer herbs sprout

Half-summer herbs (Pinellia ternata) begin to grow

30 / 72 物候
半夏生

半夏开始萌芽。

玉钩凉月挂　水麝秋蕖馥

宋·赵彦端《千秋岁》

Blue Bird Nest

Preparation : 1hr. Ingredients : ice jelly powder 15g, bird nest 50g (soaked), sago 10g, pure water 800ml. Seasonings: rock sugar 70g, butterfly pea 10g. Finishing : 60 mins. Method :1. Mix butterfly pea with 300ml water, cook sago to transparent, then drain . 2. Steam bird nest with 20g rock sugar for 20 mins. 3. Heat ice jelly powder with remaining rock sugar and water until dissolve, then add sago. 4. Pour the sago on plate and serve with the bird nest.

蓝调冰粉燕窝

准备时间：1小时。食材：冰粉粉 15 克，燕窝 50 克（泡发），西米 10 克，纯净水 800 毫升。调料：冰糖 70 克，蝶豆花 10 克。制作时间：60 分钟。做法：1. 在 300 毫升纯净水中放入蝶豆花，将西米煮至透明后沥干水分，摊凉。2. 将燕窝与 20 克冰糖蒸 20 分钟备用。3. 将 500 克纯净水与剩余冰糖及冰粉粉煮化，放入西米。4. 最后将步骤 3 倒入碗中，放上燕窝即可。

Color Combination　色彩搭配：

花带雨　　冰肌香透

宋·秦观《御街行·银烛生花如红豆》

Crispy Salty Chicken

Preparation : 15 mins. Ingredients : chicken 1pc, chicory 20g, beetroot seedling 20g, watermelon 10g (diced), spring onion 20g, ginger 10g. Seasonings : salt 20g, rock salt 20g, Sichuan pepper 5g, star anis 2g, bay leaf 1g, sugar 3g. Finishing: 12 hrs. Method : 1. Toast salt, rock salt, Sichuan pepper, star anis and bay leaf in a pan until fragrant, blend with sugar in machine. 2. Wash and dry the chicken, then rub with toast spicy salt, then keep in refrigerator overnight. 3. Boil a pot of water with spring onion and ginger, then poach the chicken for 40mins until cooked. 4. Remove the chicken into ice water, then drain. 5. Remove the bone from chicken, chop it into small pieces, plate the chicken with watermelon dice, beetroot seedling and chicory.

脆皮咸鸡

准备时间：15 分钟。食材：鲜鸡一只，苦苣叶 20 克，红菜苗 20 克，西瓜丁 20 克，小葱 20 克，姜 10 克。调料：盐 20 克，粗盐 20 克，花椒 5 克，大料 2 克，香叶 1 克，白砂糖 3 克。制作时间：12 小时。做法：1. 锅中放入盐、粗盐、花椒、大料、香叶炒香后放凉，与白砂糖放入打碎机打碎成香料盐备用。2. 鲜鸡洗净，擦干水分后用香料盐搓入味，放入冰箱冷藏一夜。3. 锅中烧开水，加入小葱与姜，放入鸡浸泡 40 分钟。4. 将鸡放入冰水中迅速过凉，沥干水分。5. 将鸡去骨切成条装盘，再用西瓜丁、红菜苗及苦苣叶点缀。

Color Combination　色彩搭配：

人生有很多唯一，过去了就再也不能重来。
——《海东妈妈做的凉面，让人心里热》

Life is filled with many one-and-only moments; once gone, they can never be revisited.
— **The Cold Noodle Made by Haidong's Mother, A Heartwarming Delight**

小暑

Lesser Heat

THE 31ST
PHENOLOGICAL PERIOD
Warm winds arrive

There's not a hint of cool breeze, only waves of heat.

31 / 72 物候
温风至

大地上不再有一丝凉风，所有的风中都带着热浪。

Ricefield Eel With Pepper In Stone Pot
Preparation : 15 mins. Ingredients : cooked yellow eel julienne 300g, onion julienne 50g, garlic slice 50g, leek 80g. Seasonings : sugar 50g, chicken stock 50ml, dark soy sauce 25 ml, light soy sauce 25 ml, cooking wine 70 ml, Zhenjiang vinegar 30 ml, ground white pepper 20g, sesame oil 50 ml. Finishing : 20 mins. Method : 1. Blanch the yellow eel julienne in boil water, drain. 2. Saute the onion julienne, garlic slice and leek in a hot wok. 3. Add the yellow eel julienne, then add sugar, chicken stock, dark soy sauce, light soy sauce, cooking wine and 5g ground white pepper, stir fry, then dress Zhenjiang vinegar when it finishes. 4. Heat the stone pot, add the yellow eel julienne and onion, serve with remaining ground white remaining pepper and sesame oil.

一碗胡椒炒软兜
准备时间：15分钟。食材：熟鳝丝300克，洋葱丝50克，蒜片50克，葱段80克。调料：白砂糖50克，清汤50毫升，老抽25毫升，生抽25毫升，料酒70毫升，镇江香醋30毫升，白胡椒粉20克，香油50毫升。制作时间：20分钟。做法：1. 熟鳝丝氽水，控干水分备用。2. 把锅烧热，爆香洋葱、蒜片与葱段。3. 放入鳝丝炒香，再放入白砂糖、清汤、老抽、生抽、料酒以及5克白胡椒粉调味，出锅时再烹入香醋。4. 将铁锅或石锅烧热，倒入香油与洋葱丝。5. 将步骤3倒入，食用时再倒入剩余的白胡椒粉即可。

鳝鱼一窜
洒水长虹
烹时五味
不妨仙骨脆

清·袁枚《红楼梦著后感二十首·自纪》

Color Combination 色彩搭配：

黄花万蕊雕阑绕　通体清香无俗调

宋·欧阳修《渔家傲》

Braised Luffa With Dried White Shrimp

Preparation : 15 mins. Ingredients : luffa 150g (peeled), dried white shrimp 10g, fried garlic 30g. Seasonings : canola seed oil 10ml, chicken stock 200ml, salt 3g, ground white pepper 1g. Finishing : 10 mins. Method : 1. Peel luffa and cut in small chunks. 2. Heat canola seed oil and fried garlic, add chicken stock, season with salt and ground white pepper. 3. Add luffa and dried white shrimp then braise for 2 mins to reduce juice, ready to serve.

芒种虾烧丝瓜

准备时间：15分钟。食材：丝瓜150克，芒种虾10克，炸蒜子30克。调料：菜籽油10毫升，鸡汤200毫升，盐3克，白胡椒粉1克。制作时间：10分钟。做法：1. 丝瓜去皮切成块备用。2. 锅中放入菜籽油与炸蒜子，倒入鸡汤并用盐和白胡椒粉调味，再放入丝瓜和芒种虾，煮2分钟收汁上盘即可。

THE 32ND PHENOLOGICAL PERIOD
Crickets seek shelter

Crickets leave the fields for cooler spots, like courtyards or gaps in stone walls.

32 / 72 物候
蟋蟀居壁

■ 蟋蟀离开了田野，到阴凉的庭院或屋角的石头缝里居住。

宋·柳永《早梅芳·海霞红》

海霞红　山烟翠

Litsea Cubeba Oil Tuna Rose

Preparation : 10 mins. Ingredients : tuna belly 300g, caviar 5g. Seasonings : litsea cubeba oil 10ml, shallot 30g (dry), celery 25g (finely chopped), galangal 10g (finely chopped), extra virgin olive oil 20ml. Finishing : 15 mins. Method : 1. Slice tuna belly then roll into rose shape. 2. Mix litsea cubeba oil, shallot, celery and galanga into dressing. Arrange the plate, top with tuna. 3. Brush extra virgin olive oil and garnish with caviar.

■

玫瑰花木姜子金枪鱼腩

准备时间：10分钟。食材：金枪鱼腩300克，鱼子酱5克。调料：木姜子油10毫升，葱末（干）30克，芹菜末25克，沙姜末10克，初榨橄榄油20毫升。制作时间：15分钟。做法：1.金枪鱼腩切薄片卷成花状，备用。2.木姜子油、葱末、芹菜末、沙姜末制成酱。垫入盘底，放上金枪鱼腩。3.金枪鱼腩上淋上少许初榨橄榄油，用鱼子酱点缀即可。

Color Combination　色彩搭配：

春草如有情　山中尚含绿

唐 · 李白《金门答苏秀才》

Poach Baby Croaker In Sichuan Pepper Oil

Preparation : 20 mins. Ingredients : baby croaker 4pcs (filleted), pickle asparagus lettuce 50g, fresh Sichuan pepper 30g. Seasonings : salt water (1:1) 800ml, Sichuan pepper paste 4g, ground white pepper 2g, Sichuan pepper oil 1l. Finishing : 20 mins. Method : 1. Soak baby croaker fillet in salt water for 1min, drain and dry. 2. Mix baby croaker fillet with ground white pepper and fresh Sichuan pepper paste. 3. Put the baby croaker fillet in a pot, add Sichuan pepper oil then heat in low fire, while the fillet turn white, drain. 4. Arrange the baby croaker fillet and pickle asparagus lettuce on plate, garnish with fresh Sichuan pepper and dress with a little of Sichuan pepper oil.

花椒油浸梅童鱼

准备时间：20分钟。食材：梅童鱼4条（取肉），泡青笋条50克，鲜花椒30克。调料：盐水（1:1）800毫升，青花椒酱4克，白胡椒粉2克，花椒油1升。制作时间：20分钟。做法：1. 梅童鱼肉用高浓度盐水爆腌1分钟，沥干水分备用。2. 将梅童鱼肉与白胡椒粉、青花椒酱拌匀。3. 锅中放入梅童鱼，倒入花椒油没过鱼肉，小火加热到鱼皮卷起后捞出，沥干油分。4. 盘中摆放泡青笋条，将鱼摆放在青笋上，再配以鲜花椒，淋少许花椒油即可。

Color Combination 色彩搭配：

THE 33RD PHENOLOGICAL PERIOD
Eagles soar high

Eagles, finding the ground too hot,
fly higher in the sky.

33 / 72 物候
鹰始鸷

老鹰因地面温度太高，飞向高空活动。

迟日江山丽　　春风花草香

唐 · 杜甫《绝句二首》

Shitake Bouquet

Preparation : 10 mins. Ingredients : shitake 20g (blanched), baby carrot 10g (blanched), baby cucumber 1pc, cherry tomato 3g, pea sprout 2g. Seasonings : Puning soy bean paste 3g, mashed potato 5g, garlic oil 5ml, extra virgin olive oil 5ml, salt 2g. Finishing : 20 mins. Method : 1. Panfry shitake with garlic oil, season with Puning soy bean paste then fill the mashed potato. 2. Toss carrot and cherry tomato with salt and extra virgin olive oil, grill by blow torch. 3. Bake shitake in 220°C oven for 5mins, then arrange the other vegetables on the shitake.

花草焗褐菇

准备时间：10分钟。食材：褐菇20克（汆水），小胡萝卜10克（汆水），黄瓜花1根，小番茄3克，豌豆尖苗2克。调料：普宁豆瓣酱3克，土豆泥5克，蒜油5毫升，初榨橄榄油5毫升，盐2克。制作时间：20分钟。做法：1.褐菇用蒜油煎香，涂抹普宁豆瓣酱后再填入土豆泥备用。2.小胡萝卜与小番茄用盐与橄榄油拌匀，用喷枪喷制出焦面。3.将褐菇放入预热220℃的烤箱中烤5分钟，再放上小胡萝卜、黄瓜花、小番茄、豌豆尖苗作装饰即可。

Color Combination　色彩搭配：

零落成泥碾作尘　只有香如故

宋·陆游《卜算子·咏梅》

Baby Aubergine With Sesame Sauce

Preparation : 30 mins. Ingredients : baby aubergine 150g, sea urchin 50g, fresh wasabi 10g (purée), garlic 2g (finely chopped). Seasonings : salt 2g, sugar 1g, light soy sauce 2ml, sesame oil 2ml, sesame paste 5g . Finishing : 40mins. Method : 1. Mix all seasonings with garlic into sesame sauce. 2. Steam baby aubergine for 15mins, cut in half and peel the meat. 3. Mix the baby aubergine meat with sesame sauce, then fill back into the skin. 4. Top with sea urchin and fresh wasabi.

海胆麻酱茄泥

准备时间：30 分钟。食材：小茄子 150 克，海胆 50 克，山葵 10 克（磨泥），蒜蓉 2 克。调料：盐 2 克，白砂糖 1 克，生抽 2 毫升，芝麻油 2 毫升，麻酱 5 克。制作时间：40 分钟。做法：1. 将蒜蓉与调料混合备用。2. 小茄子隔水蒸 15 分钟，切半后挖出肉，与调好的麻酱拌匀。3. 将拌好的茄子填回原来的茄子皮内，再放上海胆及山葵泥即可。

Color Combination　色彩搭配：

酷暑天，吃打卤面，正好读书。
——《**两吃打卤面**》

On sweltering summer days, eating braised noodles is the perfect accompaniment to reading.
— **Two Servings of Braised Noodles**

大暑

Greater Heat

THE 34TH PHENOLOGICAL PERIOD
Rotten grass turns into fireflies

Firefly larvae emerge.
Ancient people believed fireflies evolved from decaying grass.

34 / 72 物候
腐草为萤

萤火虫卵化而出，古人认为萤火虫是腐草变成。

高田二麦接山青　傍水低田绿未耕

宋 · 范成大《春日田园杂兴》

Orecchiette With White Shrimp And Prosciutto Di Parma
Preparation : 20 mins. Ingredients : white shrimp 15g (peeled), prosciutto di Parma 10g (diced), orecchiette 25g (blanched), spinach orecchiette 25g (blanched). Seasonings : salt 3g, ground white pepper 1g, chicken stock 5ml, starch 1g, water starch 5ml. Finishing: 20 mins. Method : 1. Mix salt, ground white pepper, starch and chicken stock. 2. Mix white shrimp with water starch, then quicky blanch in hot oil. Drain. 3. Fry prosciutto di Parma in low heat, add Orecchiette and white shrimp, saute with 1 .

意大利帕尔玛火腿太湖白虾仁炒猫耳朵
准备时间：20分钟。食材：白虾仁15克，意大利帕尔玛火腿丁10克，白猫耳朵25克（氽熟），菠菜猫耳朵25克（氽熟）。调料：盐3克，白胡椒粉1克，鸡汤5毫升，生粉1克，水淀粉5毫升。制作时间：20分钟。做法：1. 盐、白胡椒粉、生粉、鸡汤兑成面汁备用。2. 白虾仁与水淀粉拌匀上浆，滑油备用。3. 锅中放入意大利帕尔玛火腿丁，小火煸香，放入煮好的猫耳朵和虾仁，再放入面汁翻炒即可。

Orecchiette / Ingredients : high gluten flour 100g, water 50ml, salt 2g. Method : 1. Mix flour, water and salt into dough, then rest for 1hr. 2. Roll the dough into thin, cut into small dices. 3. Press the dices into orecchiette. P.S. : Replace water into spinach juice to make spinach orecchiette。

猫耳朵 / 食材：高筋面粉100克，水50毫升，盐2克。做法：1. 把高筋面粉、水及盐混合成面团，醒发1小时。2. 将面团擀成细条，切成细丁，再用拇指推压成螺旋形的猫耳朵。备注：制作菠菜猫耳朵将水换成菠菜汁即可。

Color Combination　色彩搭配：

Freshwater Mussel Meat Ball With Bitter Gourd

Preparation : 20 mins. Ingredients : pork belly 500g (diced), freshwater mussel 80g (sliced), bitter gourd 300g (diced). Seasonings : salt 3g, ground white pepper 3g, starch 1g, egg white 1pc, spring onion and ginger water 80ml, water starch 100ml. Finishing : 3hrs. Method : 1. Mix pork belly with spring onion and ginger water, then add egg white, salt, ground white pepper and starch, keep stirring, then add freshwater mussel, and mix until the texture become thick. 2. Make the pork belly mixture into small ball, keep in refrigerator for 30 mins. 3. Dip the meat ball into water starch, put in 60°C water to set, then cook in low heat with cover for 2hrs. 4. Add bitter gourd while it serves.

凉瓜河蚌狮子头

准备时间：20分钟。食材：五花肉 500 克（细丁），河蚌肉 80 克（切细），苦瓜 300 克（细粒）。
调料：盐 3 克，白胡椒粉 3 克，淀粉 1 克，蛋清 1 个，葱姜水 80 毫升，水淀粉 100 毫升。
制作时间：3 小时。做法：1. 五花肉丁与葱姜水搅拌，摔打上劲后加入蛋清、淀粉、盐、白胡椒粉，继续搅拌摔打，放入河蚌肉，随后挤成小球放入冷藏冰箱。2. 将定型的河蚌狮子头蘸裹一层水淀粉，再下入 60°C 的温开水中，盖盖，小火炖 2 小时。3. 最后加入苦瓜粒即可。

Color Combination 色彩搭配：

深春笋蕨千锤禄

落日渔樵三岛仙

宋·陈普《山中》

THE 35TH PHENOLOGICAL PERIOD
The earth moistens and heat intensifies

The weather becomes humid and the soil damp.

35 / 72 物候
土润溽暑

天气闷热，土地也变得潮湿。

静对西风脉脉　　金蕊绽　　粉红如滴

宋·晏殊《睿恩新》

Chill Water Melon Bun
Preparation : 10 mins. Ingredients : steamed bun 200g (break off into 15g /piece), water melon juice 500ml. Finishing : 24 hrs 15 mins. Method : 1. Dry the steamed bun in oven (60°C) for 3 hours. 2. Frozen water melon juice for 24 hours, then blend into sorbet. 3. Arrange water melon sorbet and dried bun on a chill plate.

西瓜泡馍
准备时间：10分钟。食材：戗面馒头200克（掰成15克的小块），西瓜汁500毫升。制作时间：24小时15分钟。做法：1. 将馒头放入烤箱以60℃烤3小时，烤干，备用。2. 西瓜汁放入冰箱冷冻24小时，再放入冰霜机打成冰沙。3. 将干馒头和西瓜冰霜放在冷冻过的盘上即可。

一斗擘开红玉满　　双螯啰出琼酥香

唐·唐彦谦《蟹》

Steamed Hair Crab

Preparation : 5 mins. Ingredients : hair crab 2pcs. Seasonings : ginger 50g (finely chopped), seasoned crab vinegar 50ml, brown sugar 50g, water 80ml. Finishing : 20 mins. Method : 1.Steam the hair crab for 15 mins. 2.Cook 25g ginger together with brown sugar and water, set aside. 3.Mix the remaining gingers with seasoned crab vinegar. 4.Plate the steamed crab, serve with sweet ginger water and seasoned vinegar.

六月黄

准备时间：5分钟。食材：大闸蟹2只。调料：姜末50克，蟹醋50毫升，红糖50克，水80毫升。制作时间：20分钟。做法：1.先将大闸蟹蒸15分钟。2.用25克姜末与红糖、水熬成姜糖水，备用。3.将剩余的姜末与蟹醋混合成姜醋，备用。4.将蒸好的大闸蟹上盘，食用时搭配姜糖水和姜醋即可。

Color Combination 色彩搭配：

THE 36TH PHENOLOGICAL PERIOD
Heavy rains frequent

The weather becomes humid and the soil damp.

36 / 72 物候
大雨时行

■ 雷雨也将常常出现。

风回小院庭芜绿　柳眼春相续

唐 · 李煜《虞美人 · 风回小院庭芜绿》

Chill Avocado Noodle With Sichuan Sauce
Preparation : 15 mins. Ingredients : Sichuan noodle 150g, avocado paste 25g, cucumber 15g (finely sliced), crispy fried pea 5g. Seasonings : sesame sauce 5g, ground toasted chili 2g, sugar 3g, Sichuan red oil 8ml, dark soy sauce 2ml, oyster sauce 2ml. Finishing : 15 mins. Method : 1. Mix seasonings into Sichuan noodle sauce. 2. Boil noodle for 5mins, then refresh in ice water, drain. 3. Mix noodle with avocado sauce. 4. Arrange Sichuan noodle sauce in plate, add avocado noodle, garnish with cucumber slices and crispy fried pea.

■

牛油果酱四川凉面
准备时间：15分钟。食材：四川面条150克，牛油果酱25克，黄瓜丝15克，豌豆脆5克。调料：麻酱5克，胡辣子2克，白砂糖3克，红油8毫升，老抽2毫升，蚝油2毫升。制作时间：15分钟。做法：1.将调料混合成凉面酱备用。2.面条放入沸水中煮5分钟，然后浸入冰水，控干水分备用。3.将面条与牛油果酱拌匀，卷成卷备用。4.盘中放入凉面酱与牛油果面，再配以黄瓜丝和豌豆脆即可。

Sichuan Noodle / Ingredients : high gluten flour 500g, alkaline water 20ml, cassava starch 5g, starch 50, salt 3g, egg white 1pc, spinach juice 160ml. Method : 1. Mix all ingredients into dough, rest for 1hr. 2. Roll the dough into 2mm thick, then slice into noodle by using noodle machine.

四川面条 / 食材：高筋面粉500克，碱水20毫升，木薯粉5克，淀粉50克，盐3克，蛋清1个，菠菜汁160毫升。做法：1.将食材混合揉成面团，在室温下放置1小时。2.将面团擀压成2毫米薄，用面条机制作成面条即可。

枝间新绿一重重　　小蕾深藏数点红

金 · 元好问《同儿辈赋未开海棠 · 其二》

Peony Shrimp With Bird Nest

Preparation : 8 mins. Ingredients : Peony shrimp 1pc, risotto 5g (cooked), ready to eat bird nest 4g, water-chestnut popping boba 3g, red bean popping boba 3g, black glutinous rice popping boba 3g. Seasonings : sea salt 1g, extra virgin olive oil 2ml, wasabi 1g, mayonnaise 2g, spring onion oil 1ml. Finishing : 10 mins. Method :1. Toss peony shrimp with sea salt, extra virgin olive oil and wasabi. 2. Garnish risotto, mayonnaise and spring onion oil on plate. 3. Arrange peony shrimp, water-chestnut popping boba, red bean popping boba and black glutinous rice popping boba on plate while it serves.

鲜炖燕窝配牡丹虾

准备时间：8 分钟。食材：牡丹虾 1 只，意大利米 5 克（煮熟），即食燕窝 4 克，马蹄爆珠 3 克，红豆爆珠 3 克，黑糯米爆珠 3 克。调料：海盐 1 克，特级初榨橄榄油 2 毫升，青芥末 1 克，蛋黄酱 2 克，绿葱油 1 毫升。制作时间：10 分钟。做法：1. 先将牡丹虾与海盐、橄榄油及青芥末拌匀，备用。2. 用蛋黄酱、绿葱油及意大利米在盘上画出半圆线。3. 最后将牡丹虾、马蹄爆珠、红豆爆珠、黑糯米爆珠和即食燕窝上盘即可。

Color Combination　色彩搭配：

北京人入秋要吃羊肉，除去『爆烤烧涮』羊肉，似乎『氽萝卜丝羊肉丸子』更受老百姓喜欢。
——《氽萝卜丝羊肉丸子 暖心》

— A Heart-Warming Culinary Tale: Lamb Meatballs with Shredded Radishes

Autumn in Beijing is synonymous with the warmth of lamb. Beyond the beloved roasts, grills, and hot pots, the local favorite seems to be "Lamb Meatballs with Shredded Radishes".

立秋

Beginning

of

Autumn

THE 37TH PHENOLOGICAL PERIOD
Cool winds blow

A slight chill can be felt in the wind.

37 / 72 物候
凉风至

风里有了丝丝凉意。

云随碧玉歌声转　雪绕红琼舞袖回

宋 · 晏几道《鹧鸪天 · 斗鸭池南夜不归》

Wild Rice Stem With Dried Shrimp Roe
Preparation : 10 mins. Ingredients : dried shrimp roe 10g, wild rice stem 200g (shred), chicken stock 300ml, sweet pea 30g. Seasonings : salt 5g, sugar 3g, ground white pepper 3g, canola seed oil 10ml. Finishing : 10 mins. Method : 1. Slowly heat dried shrimp roe with canola seed oil, then add chicken stock to boil. 2. Add remaining seasonings, shredded wild rice stem and sweet pea, then cook for 3 mins to serve.

虾子茭白
准备时间：10 分钟。食材：虾子 10 克，茭白 200 克（切丝），鸡汤 300 毫升，甜豆 30 克。调料：盐 5 克，白砂糖 3 克，白胡椒粉 3 克，菜籽油 10 毫升。制作时间：10 分钟。做法：1. 冷锅加热菜籽油后放入虾子，再加入鸡汤后烧开。2. 放入调料后放入茭白及甜豆，再以小火烧制 3 分钟即可。

Color Combination　色彩搭配：

更有台中牛肉炙　尚盘数脔紫光球

唐 · 李日新《题仙娥驿》

Barbecued Wagyu

Preparation : 10 mins. Ingredients : wagyu sirloin 2500g, finger lime 100g (peeled), passion fruit caviar 100g, coconut cream caviar 100g. Seasonings : barbecue sauce 1000g. Finishing : 5 hrs 30 mins. Method : 1. Slice wagyu sirloin into 3cmX12cm strips, marinate with barbecued sauce for 4hrs. 2. Roast wagyu in oven (80°C and 70% RH) for 1hr. 3. Brush the wagyu with barbecue sauce, then roast in 240 °C for another 5mins. 4. Arrange the wagyu on plate, garnish with finger lime, passion fruit caviar and coconut cream caviar.

指橙黑叉烧

准备时间：10 分钟。食材：和牛西冷牛排 2500 克，指橙 100 克（取肉），热情果鱼子酱 100 克，椰奶鱼子酱 100 克。调料：叉烧酱 1000 克。制作时间：5 小时 30 分钟。做法：1. 将牛排改刀成长 12 厘米、宽 3 厘米的长方条，用叉烧酱腌制 4 小时。2. 将腌制好的牛排放入烤箱，温度 80°C、湿度 70°C 烤 1 小时。3. 将牛排取出，刷一遍叉烧酱，再以 240°C 温度烤 5 分钟后取出。4. 将牛排上盘，再配以指橙肉、热情果鱼子酱及椰奶鱼子酱即可。

Barbecue Sauce / Ingredients: sugar 500g, preserved bean curd 20g, ground five spices 20g, hoisin sauce 240g, char siu sauce 240g, spare rib sauce 20g, sesame paste 80g, peanut butter 50g, chicken rice dark soy sauce 85ml, dark soy sauce 70ml, soy sauce 120ml, ground galanga 3g, Chinese rice sprit 10ml. Method: Mix well all the Ingredients.

叉烧汁 / 食材：白砂糖 500 克，南乳 20 克，五香粉 20 克，海鲜酱 240 克，叉烧酱 240 克，排骨酱 20 克，芝麻酱 80 克，花生酱 50 克，鸡饭酱油 85 毫升，老抽 70 毫升，酱油 120 毫升，沙姜粉 3 克，白酒 10 毫升。做法：将所有食材混合即可。

Color Combination　色彩搭配：

THE 38TH PHENOLOGICAL PERIOD
White dew descends

As days shorten and nights lengthen, dew forms in the early morning.

38 / 72 物候
白露降

昼渐短，夜渐长，水汽低温凝结，清晨的大地上出现白茫茫的露珠。

月 色 波 光 看 不 定　　玉 虹 横 卧 金 鳞 舞

宋 · 范成大《满江红》

Tempura Preserved Mandarin Fish
Preparation : 30 mins. Ingredients : preserved mandarin fish 1pc (filleted), kale 10g, caviar 20g. Seasonings : tempura flour 300g, ice water 310ml, egg yolk 2pcs, salt water 500ml, baking powder 2g. Finishing : 15 mins. Method : 1. Slice preserved mandarin fish in pieces, soak in salt water for 1min, drain. 2. Mix remaining seasonings into tempura batter, dip the preserved mandarin fish and fried in 180°C hot oil until golden color. Drain. 3. Arrange kale on plate, then place the fried preserved mandarin fish, garnish with caviar.

臭鳜鱼天妇罗配鱼子
准备时间：30 分钟。食材：臭鳜鱼 1 条（取肉），羽衣甘蓝 10 克，鱼子酱 20 克。调料：天妇罗粉 300 克，冰水 310 毫升，蛋黄 2 个，盐水 500 毫升，泡打粉 2 克。制作时间：15 分钟。做法：1. 臭鳜鱼切块，放入盐水中浸泡 1 分钟后沥干水分，备用。2. 混合剩余的调料成天妇罗炸浆，把鱼肉蘸入，再以 180°C 热油炸至金黄色捞出，沥干油。3. 盘中羽衣甘蓝垫底，放上炸好的鱼肉，用鱼子酱点缀即可。

Color Combination 色彩搭配：

冰桃碧藕脆如酥　一觞千岁母子俱

宋·杨万里《题左正卿寿慈堂》

Quick-fried Lotus With Sichuan Pepper
Preparation : 10 mins. Ingredients : crispy lotus 200g (julienne) , yellow capsicum 20g (julienne). Seasonings : salt 3g, Sichuan pepper oil 10ml, white vinegar 10ml. Finishing : 5 mins. Method : 1. Quickly blanch crispy lotus in 150℃ hot oil for 3 seconds, rinse and drain. 2. Heat Sichuan pepper oil then add crispy lotus, season with salt, add yellow capsicum, stir fry. 3. Dress with white vinegar, then ready to serve.

花椒油炒酥藕
准备时间：10分钟。食材：水果藕200克（切丝），甜黄椒20克（切丝）。调料：盐3克，花椒油10毫升，白醋10毫升。制作时间：5分钟。做法：1. 将藕丝放入150℃的油中3秒，再用水冲去浮油，沥干备用。2. 锅中烧热花椒油，放入藕丝，以盐调味后再放入黄椒丝翻炒。3. 出锅前烹入白醋，装盘即可。

THE 39TH PHENOLOGICAL PERIOD
Cicadas chirp

Cicadas that sing when sensing cold start their song.

39 / 72 物候
寒蝉鸣

感阴而鸣的寒蝉开始鸣叫。

柳 阴 直　烟 里 丝 丝 弄 碧

宋 · 周邦彦《兰陵王 · 柳阴直》

Crispy Toffee Apple

Preparation : 10 mins. Ingredients : apple 300g. Seasonings : sugar 250g, ferment batter 200g, vegetable oil 5ml. Finishing : 20 mins. Method : 1. Slice apple in small pieces. 2. Mix vegetable oil with ferment batter. 3. Dip apple into batter, deep fry in 190°C oil until golden color. 4. Heat sugar with little bit of water, until it begins caramelizing, mix it with fried apple then plate.

拔丝苹果

准备时间：10 分钟。食材：苹果 300 克。调料：白砂糖 250 克，发面 200 克，植物油 5 毫升。制作时间：20 分钟。做法：1. 苹果切椭圆形块，备用。2. 发面加植物油搅匀。3. 将苹果挂上面糊，用 190°C 油温炸至金黄色。4. 锅中放入少量水，加入白砂糖炒到拔丝的程度，放入炸好的苹果块即可。

芦叶满汀洲 寒沙带浅流

宋·刘过《唐多令·芦叶满汀洲》

Beef With Lime Zest And Sea Salt
Preparation : 5 mins. Ingredients : marbled beef 100g. Seasonings : sea salt 2g, lime zest 30g. Finishing : 5 mins. Method : 1.Chop lime zest. 2.Cook beef to medium well. Arrange on plate and sprinkle with sea salt and lime zest.

青柠海盐牛肉
准备时间：5 分钟。食材：雪花牛肉 100 克。调料：海盐 2 克，青柠檬皮 30 克。制作时间：5 分钟。做法：1. 将青柠檬皮切碎。2. 雪花牛肉煎至六成熟，撒上海盐和青柠檬皮碎装盘。

Color Combination 色彩搭配：

菜有酸甜苦辣咸，人分生旦净末丑。一个芥末，就能五味杂陈，配上五脊六兽，吃出眉眼高低。

——《北京人的味道》

In the world of food, flavors are as diverse as the people who enjoy them. Mustard, with its unique ability to encompass all flavors, is a perfect example. Whether it's paired with meat, seafood, or vegetables, it brings out the best in every dish.

—**The Taste of Beijing**

处暑

The End of Heat

219

THE 40TH
PHENOLOGICAL PERIOD
Eagles hunt birds

Eagles intensify their hunting of other birds.

40 / 72 物候
鹰乃祭鸟

老鹰开始大量捕猎鸟类。

溪梅晴照生香　　冷蕊数枝争发

宋 · 张元干《石州慢》

Chinese Yam With Preserved Plum

Preparation : 5 mins. Ingredients : Chinese yam200g (as known as yamaimo). Seasonings : blueberry jam 20g, sugar 10g, kumquat peel 5g. Finishing : 45 mins. Method : 1. Wash the Chinese yam, then steam with skin for 30 mins. 2. Peel the skin off the Chinese yam, purée, then mix well with sugar. 3. Separate half of the purée, mix well with the blueberry jam. 4. Put both purée on a spoon, then use another spoon to fold into an olive shape. 5. Garnish with kumquat peel.

话梅淮山

准备时间：5 分钟。食材：山药 200 克。调料：蓝莓果酱 20 克，白砂糖 10 克，金橘丝 5 克。制作时间：45 分钟。做法：1. 山药洗净，连皮清蒸 30 分钟。2. 等山药冷却后将皮撕去，再与白砂糖一起搅拌成泥备用。3. 把一半的山药泥与蓝莓果酱混合。4. 将步骤 2 与步骤 3 用两个勺子分别挖成橄榄状。5. 把步骤 4 上盘，点缀上金橘丝即可。

Color Combination　色彩搭配：

心事一春犹未见　余花落尽青苔院

宋 · 晏殊《蝶恋花》

Quick-fried Jelly Fish With Coriander
Preparation : 10 mins. Ingredients : jelly fish 200g (soaked), coriander 50g (finely chopped). Seasonings : rice vinegar 10ml, sesame oil 10ml, ground white pepper 4g, salt 4g. Finishing : 10 mins. Method : 1. Slice jelly fish in flat, blanch and drain. 2. Mix the seasonings into sauce. 3. Fry jelly fish in a wok, add coriander and dress sauce, stir fry for few seconds then arrange on plate.

芫爆蛰头
准备时间：10 分钟。食材：蛰头 200 克（浸泡），香菜 50 克（切末）。调料：米醋 10 毫升，芝麻油 10 毫升，白胡椒粉 4 克，盐 4 克。制作时间：10 分钟。做法：1. 蛰头片成大片后氽水沥干，备用。2. 将调料混合成汁备用。3. 锅里放入蛰头爆香，加入香菜末翻炒后倒入酱汁，翻炒装盘即可。

Color Combination　色彩搭配：

THE 41ST PHENOLOGICAL PERIOD
The world begins its descent into stillness

Everything in nature starts to wither,
emanating an air of solemnity.

41 / 72 物候
天地始肃

天地间万物开始凋零，充满肃杀之气。

稠 红 乱 蕊　漫 开 遍　楚 江 南 北

宋 · 曾纡《上林春》

Osmanthus Cake
Preparation : 10 mins. Ingredients : glutinous rice flour 200g, rice 80g. Seasonings : sugar 30g, red yeast rice 2g, osmanthus wine 3ml, red bean paste for filling 30g. Finishing : 30 mins. Method : 1. Combine and mix glutinous rice flour, rice, sugar, red yeast rice, and osmanthus wine with water to form dough. 2. Stuff dough with red bean paste, place in mould and steam for 15 minutes.

桂花糕
准备时间：10 分钟。食材：糯米粉 200 克，大米 80 克。调料：白砂糖 30 克，红曲米粉 2 克，桂花酒 3 毫升，豆沙馅 30 克。制作时间：30 分钟。做法：1. 将糯米粉、大米与白砂糖、红曲米粉、桂花酒、水混合。2. 和面，包入豆沙馅，置入模具中蒸 15 分钟即可。

风雨晴时春已空　　谁惜泥沙万点红

宋 · 陆游《豆叶黄》

Baked Sole With Red Pepper

Preparation : 30 mins. Ingredients : sole 1pc (filleted), garlic 200g, diced ginger 200g, diced leek 200g. Seasonings : salt 4g, red pepper corn 2g, white pepper corn 4g (crushed), rice spirits 20ml, salt water 1500ml (1:1), peanut oil 10ml. Finishing : 20 mins. Method : 1. Soak sole fillet into salt water, drain and slice in pieces, mix with salt and white pepper. 2. Heat peanut oil in clay pot, then add garlic, ginger and leek to fry. 3. Arrange sole fillet into the clay pot, cover and cook for 6mins. 4. Dress with red pepper corn and sprinkle rice spirits.

红胡椒焗比目鱼

准备时间：30 分钟。食材：比目鱼 1 条（起肉），蒜子 200 克，姜块 200 克，大葱段 200 克。调料：盐 4 克，红胡椒粒 2 克，白胡椒碎 4 克，米酒 20 毫升，盐水 1500 毫升（1:1），花生油 10 毫升。制作时间：20 分钟。做法：1. 比目鱼浸入盐水中浸泡 1 分钟，取出沥干水分后切成小块，与盐、白胡椒碎拌匀备用。2. 砂锅中放入花生油，放入蒜子、姜块、大葱段。3. 码放上比目鱼，盖盖焗 6 分钟。4. 最后撒入红胡椒粒，喷入米酒即可。

Color Combination　色彩搭配：

THE 42ND PHENOLOGICAL PERIOD
Crops mature

It's harvest time for crops.

42 / 72 物候

禾乃登

农作物成熟。

愿君多采撷
此物最相思

唐·王维《相思》

Red Bean Soup With Bird Nest

Preparation : 12hrs. Ingredients : bird nest 50g (soaked), red bean 230g (soaked overnight). Seasonings : brown sugar 30g, rock sugar 30g, dried orange peel 30g (soaked), water 1000ml. Finishing : 3.5hrs. Method : 1. Steam bird nest for 20mins. 2. Steam red bean with water for 2hrs, pass sieve to make purée. 3. Add red bean purée, brown sugar and rock sugar in pot, heat for 1hr in low fire, then add dried orange peel. 4. Add red bean soup in a bowl, then add bird nest.

红豆沙燕窝

准备时间：12小时。食材：燕窝 50 克（泡发），红豆 230 克（隔夜浸泡）。调料：红糖 30 克，冰糖 30 克，陈皮 30 克，水 1000 毫升。制作时间：3.5 小时。做法：1. 燕窝隔水蒸 20 分钟备用。2. 红豆加水蒸制 2 小时，过筛成红豆沙。3. 锅中放入红豆、红糖、冰糖，小火熬 1 小时后加入陈皮，保温备用。4. 碗中放入红豆沙，再放入燕窝即可。

Color Combination　色彩搭配：

炉火照天地　红星乱紫烟

唐 · 李白《秋浦歌十七首 · 其十四》

Shrimp With Pickle
Preparation : 10 mins. Ingredients : lamb lettuce 5g, shrimp 20g, purple cauliflower 10g, avocado 20g (sliced). Seasonings : pickle juice 300ml, black pepper 3g (crushed), corn starch 3g, water 10ml. Finishing : 20 mins. Method : 1. Toss shrimp with corn starch and water, poach in warm oil. Drain. 2. Fry shrimp in wok, dress 5ml pickle juice, set aside. 3. Break purple cauliflower in small pieces, soak in pickle juice for 10mins. 4. Arrange shrimp, purple cauliflower, avocado and lamb lettuce on plate, garnish with black pepper.

泡菜虾仁
准备时间：10 分钟。食材：鸡毛菜 5 克，虾仁 20 克，紫菜花 10 克，牛油果 20 克（切片）。调料：泡菜汤 300 毫升，黑胡椒碎 3 克，玉米淀粉 3 克，水 10 毫升。制作时间：20 分钟。做法：1. 虾仁与玉米淀粉、水混合上浆，滑油沥干。2. 虾仁入锅快炒，喷入 5 毫升泡菜汤后备用。3. 紫菜花掰小朵，放入泡菜汤中浸泡 10 分钟，沥干备用。4. 将虾仁、紫菜花、牛油果及鸡毛菜上盘，撒上黑胡椒碎即可。

Color Combination　色彩搭配：

我爱北京的秋，怀念霞帔中烤鸭果木烧烤的味儿。

——《北京『三烤』烤鸭》

I love autumn in Beijing, especially reminiscing about the aroma of Peking duck roasted over fruitwood.

—Beijing's 'Triple Roast': Peking Duck

白露

White Dew

THE 43RD PHENOLOGICAL PERIOD
Wild geese arrive

Large and small geese begin their migration to the south for winter.

43 / 72 物候
鸿雁来

大曰鸿，小曰雁，鸟儿就要开始飞往南方去过冬了。

常时黄色见眉间　松桂我同攀

宋 · 王安石《诉衷情 · 和俞秀老鹤词 · 五之一》

Pineapple Pickle

Preparation : 5 mins. Ingredients : pineapple 120g, confit orange zest 12g, coconut caviar 6g. Seasonings : pickle water 300ml. Finishing : 45 mins. Method :1.Slice pineapple into 4.5cm X 2cm X1.5cm pieces, then soak in pickle water for 30mins. 2. Dry pineapple then arrange on plate, garnish with confit orange zest and coconut caviar.

老坛泡凤梨

准备时间：5 分钟。食材：凤梨肉 120 克，糖渍橙皮丝 12 克，椰浆鱼籽 6 克。调料：泡菜汁 300 毫升。制作时间：45 分钟。做法：1. 凤梨切成长 4.5 厘米、宽 2 厘米、高 1.5 厘米的长条，放入泡菜汁中浸泡半小时备用。2. 将泡好的凤梨沥干水分，再点缀上糖渍橙皮丝及椰浆鱼籽即可。

Confit Orange Zest / Ingredients: orange zest 100g, sugar 300g, water200ml. Method: 1. Boil sugar and water into syrup, when it cold down to 60℃ , add orange zest to marinade for 24hrs. 2. Wash and slice while use.

糖渍橙皮丝 / 食材：橙皮 100 克，白砂糖 300 克，水 200 毫升。做法：1. 白砂糖与水煮成糖浆，待糖浆凉至 60℃时放入橙皮腌渍 24 小时。2. 使用时洗净切丝即可。

Pickle Water / Ingredients: purify water 5l, leek 100g, ginger 80g, bird eye chili 70g, amomum tsao-ko 5pcs, star anise 5pcs, Sichuan pepper 40g, pickle chili 250g, salt 350g, white vinegar750ml, sugar 15g, Beijing white liquor 20ml. Method: Mix all the ingredients, pour into a container, and ferment at room temperature for three days.

泡菜汁 / 食材：纯净水 5 升，大葱 100 克，姜 80 克，小米辣 70 克，草果 5 颗，八角 5 颗，鲜花椒 40 克，小米辣泡椒 250 克，盐 350 克，白醋 750 毫升，白砂糖 15 克，二锅头 20 毫升。做法：将所有材料混合，倒入容器中，室温下发酵三天即可。

Color Combination　色彩搭配：

十里楼台倚翠微　百花深处杜鹃啼

宋 · 晏几道《鹧鸪天 · 十里楼台倚翠微》

Geoduck With Fried Yunna Mushroom

Preparation : 10 mins. Ingredients : geoduck 1pc (peeled). Seasonings : Maggi 3ml, steam fish soy sauce 8ml, fried Yunna mushroom 30g. Finishing : 10 mins. Method : 1. Slice geoduck in thin, quickly blanch in boiling water, refresh in ice water, drain. 2. Mix all the seasonings, then arrange with geoduck on plate.

加蚌油鸡枞

准备时间：10分钟。食材：象拔蚌1只（取肉去皮）。调料：美极鲜3毫升，蒸鱼豉油8毫升，油鸡枞30克。制作时间：10分钟。做法：1.象拔蚌片成薄片，汆水后浸入冰水，控干水分。2.美极鲜、蒸鱼豉油与油鸡枞拌匀，与象拔蚌片一起装盘即可。

Color Combination　色彩搭配：

THE 44TH PHENOLOGICAL PERIOD
Swallows depart

Swallows and other migratory birds head south to avoid the cold.

44 / 72 物候
玄鸟归

燕子等候鸟也要南飞避寒了。

剪翠妆红欲就
折得清香满袖

宋·晏殊《雨中花》

Marinated Fennel
Preparation : 10 mins. Ingredients : fennel 1pc, bird eye chili 10g (sliced). Seasonings : salt 5g, marinated sauce 600ml. Finishing : 2 hrs. Method : 1. Cut fennel in thick, marinate with salt for 30mins, flipping and mixing every 15 mins, rise, then soak in marinated sauce for 1hr. 2. Drain and dry fennel, arrange with chili on plate.

香辣茴香根
准备时间：10 分钟。食材：茴香根 1 个，朝天椒 10 克（切小粒）。调料：盐 5 克，调味汁 600 毫升。制作时间：2 小时。做法：1. 先将茴香根切片，用盐腌半小时（每 15 分钟翻拌一下），冲水，再用调味汁腌制 1 小时以上，入味即可。2. 把茴香根沥干水分，与朝天椒粒一起上盘即可。

Marinated Sauce
Ingredients: Tabasco 60g, oyster sauce 150ml, Maggi 50ml, soy sauce 550ml, dark sugar syrup 50ml, rock sugar 250g, sugar 1200g, white vinegar 500ml, rice vinegar 120ml, sliced garlic 50g, chili 40g, coriander 80g, purified water 200ml, salt 8g, star anis 1pc, cinnamon 1g, bay leaf 2pcs. Method: Put all the ingredients in a pot, boil over low heat, reserve for cooling.

腌茴香根调味汁
食材：TABASCO 辣椒仔酱 60 克，蚝油 150 毫升，美极鲜 50 毫升，酱油 550 毫升，红糖水 50 毫升，冰糖 250 克，白砂糖 1200 克，白醋 500 毫升，米醋 120 毫升，蒜片 50 克，小米椒 40 克，香菜 80 克，纯净水 200 毫升，盐 8 克，大料 1 块，桂皮 1 克，香叶 2 片。做法：将所有食材放入锅中以小火熬煮，摊凉备用。

Color Combination 色彩搭配：

应是无机承雨露
却将春色寄苔痕

唐·长孙佐辅《拟古咏河边枯树》

Assort Vegetables With Truffles
Preparation : 10 mins. Ingredients : porcini 75g (blanched), button mushroom 75g (blanched), Chinese yam 25g, chestnut 25g. Seasonings : butter 10g, chicken stock 50ml, truffle sauce 3g, truffle oil 2ml, salt 1g, sugar 2g, oyster sauce 3ml. Finishing : 15 mins. Method : 1. Pan-fry porcini, button mushroom and Chinese yam with butter. 2. Fry truffle sauce, add porcini, button mushroom and chestnut, add chicken stock and seasonings. 3. Arrange all the vegetables on plate, dress with truffle oil.

松露全素
准备时间：10分钟。食材：牛肝菌75克（氽水），白口蘑75克（氽水），山药25克，栗子25克。调料：黄油10克，鸡汤50毫升，松露酱3克，松露油2毫升，盐1克，白砂糖2克，蚝油3毫升。制作时间：15分钟。做法：1. 牛肝菌、白口蘑及山药用黄油煎香备用。2. 锅中放入松露酱煸出香气，放入牛肝菌、白口蘑、栗子，加入鸡汤及剩余调料后小火收汁上盘。3. 最后放上煎山药，滴上松露油即可。

Color Combination 色彩搭配：

THE 45TH PHENOLOGICAL PERIOD
Birds gather food

Birds start storing food in preparation for winter.

45 / 72 物候
群鸟养羞

■ 百鸟开始贮存粮食以备过冬。

灯火钱塘三五夜　明月如霜　照见人如画

宋 · 苏轼《蝶恋花 · 密州上元》

Tofu Pudding With Ginger Syrup

Preparation : 90 mins. Ingredients : milk 215ml, tofu 65g, ginger juice 50ml, dried pea 10g (soaked). Seasonings : sugar 10g, brown sugar 50g, water 50ml, gelatine sheet 1pc (soaked). Finishing : 60 mins. Method : 1. Blend milk, sugar and tofu, heat with gelatine sheet, then pour into a mould, keep in refrigerator to set. 2. Steam pea for 20mins until soft, drain. 3. Mix ginger juice, water and brown sugar, then reduce until syrup thicken. 4. Plate the tofu pudding on plate, dress ginger syrup and steamed pea.

■

姜汁红糖豆腐冻

准备时间：90 分钟。食材：牛奶 215 毫升，豆腐 65 克，姜汁 50 毫升，去皮豌豆 10 克（浸泡）。调料：白砂糖 10 克，红糖 50 克，水 50 毫升，鱼胶片 1 片（泡软）。制作时间：60 分钟。做法：1. 将牛奶、白砂糖、豆腐打碎，加热后放入鱼胶片，倒入模具后，放入冰箱冷藏凝固。2. 去皮豌豆隔水蒸 20 分钟，备用。3. 姜汁和水混合，加入红糖熬稠晾凉。4. 将凝固的豆腐冻放入盘中，淋上姜汁、红糖浆，撒上豌豆即可。

泪眼问花花不语
乱红飞过秋千去

宋·欧阳修《蝶恋花·庭院深深几许》

Almond Ice-Cream

Preparation : 20 mins. Ingredients : almond 100g, water 550ml, sago 100g, rose caviar 50g. Seasonings : sugar 130g. Finishing : 13 hrs. Method : 1. Soak almond with water overnight, blend and pass sieve to extract almond mlik. 2. Heat almond milk with sugar in low heat, let cool, then pour into pacojet can, frozen for 12hrs. 3. Boil sago for 8mins, refresh in ice water, drain. 4. Use parcojet to make almond ice-cream, scoop to make olive shape, then place on plate. 5. Garnish with rose caviar and sago.

杏汁冰淇淋

准备时间：20分钟。食材：杏仁100克，水550毫升，西米100克，玫瑰鱼子50克。调料：白砂糖130克。制作时间：13小时。做法：1. 杏仁用水隔夜浸泡，然后连水一起打碎后过纱布滤出杏汁。2. 将杏汁与白砂糖用小火烧开，晾凉后装入pacojet罐中，放入冰箱冷冻12小时。3. 将西米放入沸水中煮8分钟，捞出放入冰水中镇凉，沥干水分。4. 将冷冻好的杏汁用pacojet罐打至顺滑，用勺子挖成橄榄形上盘。5. 最后配以西米及玫瑰鱼子即可。

秋天帝王蟹特别肥，口感水嫩水嫩的，鲜甜。尤其是回味儿。

——《帝王蟹的味道》

In autumn, the king crab is especially plump, tender, and sweet, particularly the aftertaste.

— **The Flavor of King Crab**

秋分

Autumnal

Equinox

THE 46TH PHENOLOGICAL PERIOD
Thunder recedes

46 / 72 物候
雷始收声

Post-autumn equinox, thunderstorms become rare.

秋分之后便不再打雷。

旧时王谢堂前燕
飞入寻常百姓家

唐 · 刘禹锡《乌衣巷》

Pumpkin Soup With Bird Nest

Preparation : 15 mins. Ingredients : pumpkin 230g, ready to eat bird nest 8g, pumpkin flower 1pc, shallot 20g (finely chopped), Parmigiano Reggiano 2g (grinded). Seasonings : chicken stock 100ml, curry powder 5g, fried batter 200g, salt 5g, cream 30g, butter 20g. Finishing : 50 mins. Method : 1. Dip pumpkin flower into fried batter, deep fry until golden color, drain. 2. Scoop pumpkin into 1cm ball, blanch and drain. Then cut remaining pumpkin into small pieces. 3. Sweat shallot with butter, add pumpkin stir-fry, add chicken stock and curry powder, cook in low heat for 15mins. 4. Add cream and salt, then blend into cream soup. 5. Arrange Parmigiano Reggiano, pumpkin ball, fried pumpkin flower and bird nest on plate, pour pumpkin soup when it serves.

燕窝南瓜花汤

准备时间：15 分钟。食材：南瓜 230 克，即食燕窝 8 克，南瓜花 1 朵，红葱 20 克（切碎），帕马森奶酪 2 克 (磨碎)。调料：鸡汤 100 毫升，咖喱粉 5 克，脆炸糊 200 克，盐 5 克，奶油 30 克，黄油 20 克。制作时间：50 分钟。做法：1. 南瓜花用水冲洗，沥干水分，里外均抹上薄薄一层脆炸糊，入锅炸至金黄酥脆，备用。2. 南瓜挖出数个直径约 1 厘米的小球，每个约 15 克，余水沥干。将剩余的南瓜切小块，备用。3. 将红葱用黄油小火煸香，加入南瓜块小火煸炒，再倒入鸡汤及咖喱粉，煮 15 分钟。4. 加入奶油与盐后，用破壁机打成浓汤。5. 盘中撒上帕马森奶酪碎、南瓜球、炸南瓜花、即食燕窝，食用时浇上南瓜浓汤。

Fried batter / Ingredients: tempura powder 120g, cornstarch 40g, ice water 200 ml. Method : Mix well all the ingredients.

脆炸糊 / 食材：天妇罗粉 120 克，玉米淀粉 40 克，冰水 200 毫升。做法：将食材混合均匀即可。

Color Combination 色彩搭配：

Daikon Soup With Mini Lamb Ball

Preparation : 20 mins. Ingredients : minced lamb meat 500g, daikon 500g (shredded). Seasonings : sesame oil 5ml, salt 6g, ground white pepper 6g, ginger and spring onion water 120ml, egg white 10g, corn starch 15g, flour 5g. Finishing : 45 mins. Method : 1. Beat lamb with ginger and spring onion water, season with salt and ground white pepper, then add egg white, flour and corn starch, keep stirring until texture become hard. Keep in refrigerator. 2. Heat a pot of water to 90°C, press the lamb into small ball and poach in water, add sesame oil. 3. Add shredded daikon, poach for another 15 mins to serve.

寒羊肉如膏
江鱼如切玉

宋·张耒《冬日放言二十一首其一》

■ **萝卜羊肉氽丸子**

准备时间：20 分钟。食材：羊肉馅 500 克，白萝卜 500 克（刨丝）。调料：芝麻油 5 毫升，盐 6 克，白胡椒粉 6 克，葱姜水 120 毫升，鸡蛋清 10 克，玉米淀粉 15 克，面粉 5 克。制作时间：45 分钟。做法：1. 羊肉馅与葱姜水搅拌，加入盐、白胡椒粉、鸡蛋清、面粉、玉米淀粉继续搅拌，放入冰箱冷藏备用。2. 锅中将水加热至 90℃，将羊肉馅挤成小丸子下锅，再放入芝麻油。3. 放入白萝卜丝，以小火慢炖 15 分钟即可。

THE 47TH
PHENOLOGICAL PERIOD
Hibernating insects
seal their homes

Insects that hibernate start to
seal their burrows to keep out the cold.

47 / 72 物候
蛰虫坯户

蛰居的小虫开始藏入穴中,并且用细土将洞口封起来以防寒气侵入。

黄 花 醉 了　　碧 梧 题 罢　　闲 卧 对 高 秋

宋 · 晏几道《少年游 · 雕梁燕去》

Fish In Crab Style
Preparation : 5 mins. Ingredients : egg 1pc, mandarin fish fillet 10g, some chrysanthemum. Seasonings : salt 5g, rice vinegar 20ml, ginger juice 5ml. Finishing : 25 mins. Method : 1. Steam fish fillet to cook, mix with egg. 2. Season with salt and ginger juice, saute to cook. Dress rice vinegar. 3. Garnish with chrysanthemum.

赛螃蟹
准备时间:5 分钟。食材:鸡蛋 1 个,鳜鱼肉片 10 克,菊花少许。调料:盐 5 克,米醋 20 毫升,姜汁 5 毫升。制作时间:25 分钟。做法:1. 将蒸熟的鱼片放入打散的鸡蛋里。2. 加入盐、姜汁,上锅炒熟,最后倒入米醋调味。3. 上盘时配菊花装饰即可。

Color Combination　色彩搭配:

最宜红烛下
偏称落花前

唐·刘禹锡《抛球乐词》

Hawthorn Raspberry
Preparation : 5 mins. Ingredients : raspberry 30g, hawthorn jelly 80g. Seasonings : gelatine sheet 4pcs (soaked), grenadine 500ml. Finishing : 80 mins. Method : 1. Heat grenadine and gelatine sheet, then pour in flat sheet pan, cover then keep in refrigerator to set. 2. Scoop hawthorn jelly into small ball, arrange together with raspberry on plate, cover with grenadine jelly.

山楂覆盆子
准备时间：5 分钟。食材：覆盆子 30 克，山楂糕 80 克。调料：鱼胶片 4 片（泡软），石榴糖浆 500 毫升。制作时间：80 分钟。做法：1. 将石榴糖浆与鱼胶片烧开后倒入平盘，包上保鲜膜放入冰箱冷藏直至凝固成啫喱片。2. 山楂糕挖球，并与覆盆子一起放入盘中，覆盖上啫喱片即可。

Color Combination　色彩搭配：

THE 48TH PHENOLOGICAL PERIOD
Water starts to dry up

After the rainy spring and summer, moisture begins to deplete.

48 / 72 物候
水始涸

■ 春夏期间雨水多，空气湿润，到秋冬便渐渐干涸，水汽渐消。

宋·方岳《采菌》

石鼎香漠漠
甘餐自当肉

Stewed Matsutake With Spring Water
Preparation : 5 mins. Ingredients : fresh matsutake 50g, natural spring water 100g. Seasonings : salt 0.5gl. Finishing : 25 mins. Method : 1. Wash matsutake, slice in thick, then put in a bowl. 2. Boil spring water, add salt. 3. Pour the spring water into bowl, then steam for 20 mins.

■

泉水松茸
准备时间：5 分钟。食材：鲜松茸 50 克，矿泉水 100 毫升。调料：盐 0.5 克。制作时间：25 分钟。做法：1. 松茸洗净切片，放入碗中备用。2. 矿泉水烧开后加入盐，再倒入松茸碗中。3. 将松茸上锅隔水蒸 20 分钟即可。

Color Combination　色彩搭配：

白纻流吴曲　红花烂楚芳

元 · 王冕《自崔镇至济州人情风俗可叹三十韵》

Steam Glutinous Rice Cake With Maple Syrup Cherry

Preparation : 12hrs. Ingredients : glutinous rice 500g (soaked), dried fig 50g (soaked), dried date500g (peeled), red bean purée 300g, cherry 500g (peeled). Seasonings : water20ml, alkaline water 2ml, maple syrup 50ml. Finishing : 90 mins. Method : 1. Heat maple syrup to 60℃ , then mix with cherry. Keep cool. 2. Steam glutinous rice for 30mins, rinse with cold water, drain. 3. Mix alkaline water with date, steam fig to soft, cut in half. 4. Divide glutinous rice in 4 portions, then layer date, red bean purée, glutinous rice, fig in a bowl. Steam for 30mins. 5. Top with cherry and ready to serve.

枫糖浆渍野樱桃甑糕

准备时间：12 小时。食材：糯米 500 克（浸泡），无花果干 50 克（泡软），蜜枣 500 克（去核），红豆馅 300 克，樱桃 500 克（去核）。调料：水 20 毫升，碱水 2 毫升，枫糖浆 50 毫升。制作时间：90 分钟。做法：1. 枫糖浆加热至 60℃后与樱桃拌匀，自然冷却。2. 糯米隔水蒸 30 分钟，再用凉水冲散，沥干水分备用。3. 蜜枣与碱水混合备用，无花果干蒸软，切开备用。4. 将蒸好的糯米分成 4 份，按顺序将蜜枣、红豆馅、糯米、无花果干装入碗中，再隔水蒸 30 分钟。5. 最后将蜜制好的樱桃放上即可。

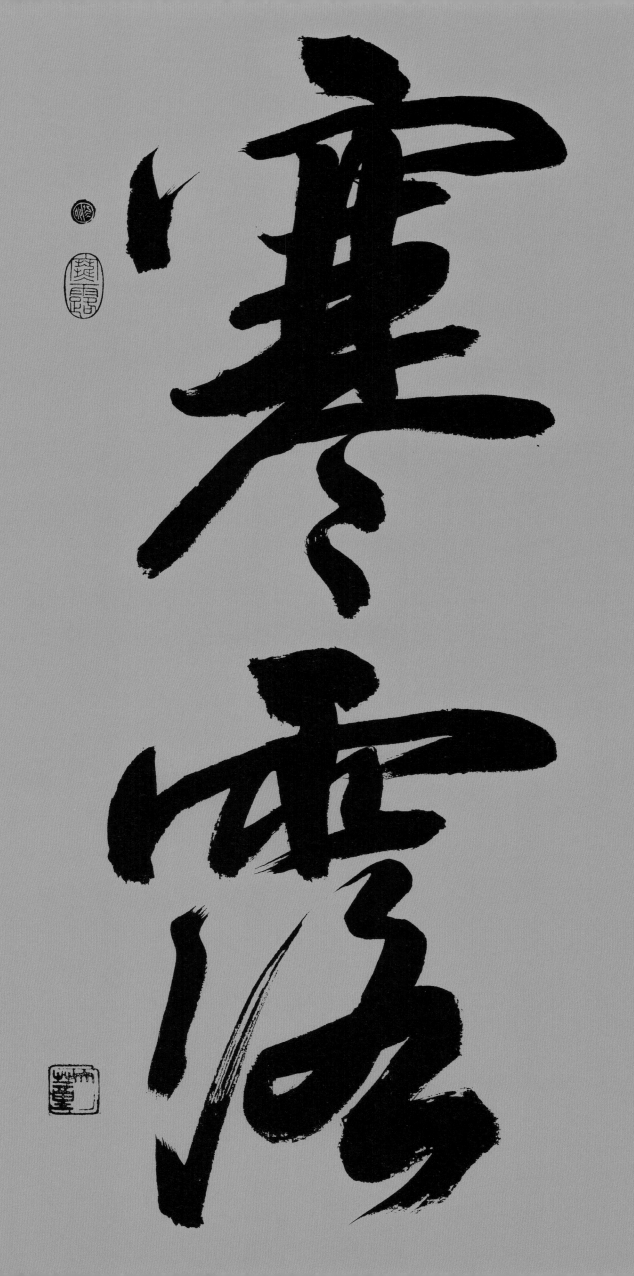

秋天吃花椒，宜老友，但说新话，兴致渐浓而不悲秋。

——《吃花椒宜老友说新话，不悲秋》

Autumn is the perfect season for indulging in Sichuan peppercorns with old friends. It's a time for sharing new stories, and creating a lively atmosphere that defies the melancholy of the season.

—*Autumnal Gatherings: Sichuan Peppercorns, Old Friends, and New Stories*

寒露

Cold Dew

THE 49TH PHENOLOGICAL PERIOD
Wild geese are honored guests

Geese fly south in V-shaped formations.

49 / 72 物候
鸿雁来宾

鸿雁排成一字或人字形的队列大举南迁。

凡物从来遇合难　烂甜中故有微酸

宋 · 苏洞《悼杨梅》

Hawthorn With Rose Caviar
Preparation : 10 mins. Ingredients : hawthorn 280g, rose caviar 20g. Seasonings : lemon juice 10ml, sugar 250g, pectin 10g. Finishing : 4 hrs. Method : 1. Blend hawthorn into purée, pass sieve. 2. Cook hawthorn with sugar, lemon juice and pectin until thick. 3. Pour hawthorn into 2.5×2.5cm mould, keep in refrigerator for 3 hours until set. 4. Arrange the hawthorn on plate, then garnish with rose caviar.

山楂糕和玫瑰鱼子
准备时间：10 分钟。食材：山楂 280 克，玫瑰鱼子 20 克。调料：柠檬汁 10 毫升，白砂糖 250 克，果胶 10 克。制作时间：4 小时。做法：1. 山楂打碎过筛，与白砂糖、柠檬汁、果胶一同放入锅中熬至黏稠。2. 将熬制好的山楂倒入 2.5×2.5 厘米的模具中，放入冰箱冷藏 3 小时定型。3. 将山楂糕上盘，点缀上玫瑰鱼子。

Color Combination 色彩搭配：

主人留客醉
酒美蟹螯肥

宋·滕元发《句·其六》

Alaska King Crab Curry

Preparation : 20 mins. Ingredients : Alaska king crab 1pc（2.5kg）, red capsicum 100g (thin strip), green capsicum 100g (thin strip), onion 100g (thin strip). Seasonings : curry sauce 800ml, chili oil 50ml. Finishing : 45 mins. Method : 1. Steam Alaska king crab, peel the meat. 2. Fry red capsicum, green capsicum, onion in 200 °C for 2mins, drain. 3. Boil curry sauce in low heat, add Alaska king crab, red capsicum, green capsicum, onion, stir and arrange on plate. 4. Dress chili oil and cover with Alaska king crab shell. P.S.: Curry sauce, please see the recipe of Alaska king crab hot pot (P370).

咖喱阿拉斯加蟹

准备时间：20分钟。食材：阿拉斯加蟹1只(约2.5公斤)，甜红椒100克（切条），甜绿椒100克（切条），洋葱100克（切条）。调料：咖喱汁800毫升，辣椒油50毫升。制作时间：45分钟。做法：1.阿拉斯加蟹蒸熟，去壳，取肉备用。2. 将甜红椒、甜绿椒、洋葱用200℃油温的食用油炸2分钟，沥干备用。3. 将咖喱汁小火烧开，放入阿拉斯加蟹、甜红椒、甜绿椒、洋葱，拌匀上盘。4. 最后淋上辣椒油，盖上蟹盖即可。备注：咖喱汁制作请参考涮阿拉斯加蟹（P370）。

Color Combination 色彩搭配：

THE 50TH PHENOLOGICAL PERIOD
Sparrows turn into clams

As the cold deepens, sparrows vanish.
Ancient people, seeing the sudden appearance of many clams by the sea,
believed clams were transformed sparrows.

50 / 72 物候
雀入大水为蛤

深秋天寒，雀鸟都不见了，古人看到海边突然出现很多蛤蜊，认为蛤蜊是雀鸟变成的。

春阴垂野草青青　　时有幽花一树明

宋 · 苏舜钦《淮中晚泊犊头》

Dry Braised Squab With Lemon Grass
Preparation : 5 mins. Ingredients : whole squab 1pc. Seasonings : lemon grass 300g, water 350ml, salt 3g, sugar 2g. Finishing : 13 hrs. Method : 1. Blend lemon grass, season with sugar, salt. 2. Clean and trim squab, then marinate in seasoned lemon grass for 12 hours. 3. Bake squab with lemon grass at 180°C for 15 minutes.

香茅焗乳鸽
准备时间：5分钟。食材：乳鸽1只。调料：香茅300克，水350毫升，盐3克，白砂糖2克。制作时间：13小时。做法：1.香茅加水榨汁，用盐、白砂糖调味。 2.乳鸽净膛洗净，放入香茅汁中浸泡12小时。 3.将乳鸽放入预热至180°C的烤箱，烤15分钟即可。

试问甜言软语　何如大醉高吟

宋 · 郭应祥《西江月 · 妙句春云多态》

Risotto With Truffle Oil And Braised Abalone

Preparation : 20 mins. Ingredients : abalone 1pc, risotto 100g, preserved vegetables15g (meigancai, soaked). Seasonings: broth 150ml, oyster sauce 3ml, dark soy sauce 2ml, sugar 1g, truffle oil 15ml, salt 2g, cream 80ml. Finishing : 12 hrs. Method : 1.Cook risotto with water and cream, dash with truffle oil. 2.Braise abalone with preserved vegetables (meigan cai) in broth until tender, season with oyster sauce, dark soy sauce, sugar and salt. 3. Fill plate with risotto topped with 2.

松露汁烧鲍鱼配意大利米

准备时间：20 分钟。食材：鲍鱼 1 只，意大利米 100 克，梅干菜 15 克（泡发）。调料：高汤 150 毫升，蚝油 3 毫升，老抽 2 毫升，白砂糖 1 克，松露油 15 毫升，盐 2 克，淡奶油 80 毫升。制作时间：12 小时。做法：1. 意大利米加水和淡奶油煨熟，拌入松露油。2. 将鲍鱼用高汤烧透，加入梅干菜并用蚝油、老抽、白砂糖、盐调味。3. 用意大利米垫底，上面码放梅干菜鲍鱼即可。

Color Combination　色彩搭配：

THE 51ST PHENOLOGICAL PERIOD
Chrysanthemums bloom

Chrysanthemums gradually bloom, thriving in autumn.

51 / 72 物候
菊有黄花

■ 菊花渐次开放，独盛于秋。

堂前种山丹　　错落马脑盘

宋 · 苏轼《次韵子由所居六咏》

Bake Lily Bud With Jamon Iberico
Preparation : 20 mins. Ingredients : lily bud 500g (6pcs), shallot 80g (sliced), ginger 10g (sliced), jamon iberico 5g (diced). Seasonings : lily seasoning sauce 80ml, soup-stock 100ml, butter 30g, lard 30g. Finishing : 20 mins. Method : 1. Sweat shallot and ginger with butter and lard, add lily bud and remaining seasonings, cover in middle heat for 10 mins. 2. Dress diced jamon iberico, and ready to serve.

■

伊比利亚火腿焗百合
准备时间：20分钟。食材：百合球500克，红葱80克（切碎），生姜10克（切碎），伊比利亚火腿5克（切碎）。调料：百合汁80毫升，高汤100毫升，黄油30克，猪油30克。制作时间：20分钟。做法：1. 红葱末米与姜米用黄油及猪油煸香，放入百合汁和高汤，盖上盖以中火烧10分钟。2. 待汁收浓，撒上火腿粒即可。

Lily Seasoning Sauce /Ingredients: soy sauce for seafood 15ml, honey 6g, oyster sauce 10ml, cooking wine 20ml, soy sauce for chicken rice 10ml. Method: Mix well all ingredients into sauce.

百合调味汁 / 食材：蒸鱼豉油15毫升，蜂蜜6克，蚝油10毫升，料酒20毫升，鸡饭老抽10毫升。做法：将食材混合均匀即可。

日上花梢　莺穿柳带　犹压香衾卧

宋 · 柳永《定风波 · 自春来》

Braised Fish Maw With Red Vinegar Broth
Preparation : 10 hrs. Ingredients : fish maw 200g (soaked), sago 5g (cooked), caviar 3g, fresh lily bulb 10g (blanched). Seasonings : red vinegar 15ml, beetroot juice 3ml, salt 2g, sugar 1g, chicken stock 800ml, cream 2ml, water starch 5ml. Finishing : 25 mins. Method : 1. Heat chicken stock with red vinegar and beetroot juice, season with salt and sugar. 2. Add fish maw and braise for 20 mins in low heat, thicken with water starch, then add cream. 3. Place sago and lily bulb in a bowl, arrange fish maw and soup. 4. Garnish with caviar.

大红浙醋烧花胶
准备时间：10 小时。食材：花胶 200 克（浸发），西米 5 克（煮熟），鱼子酱 3 克，百合 10 克（汆水）。调料：大红浙醋 15 毫升，紫菜头汁 3 毫升，盐 2 克，白砂糖 1 克，鸡汤 800 毫升，淡奶油 2 毫升，水淀粉 5 毫升。制作时间：25 分钟。做法：1. 鸡汤加大红浙醋、紫菜头汁烧开，以盐、白砂糖调味。2. 放入花胶，小火烧制 20 分钟，放入水淀粉苟芡，再加入淡奶油。3. 将西米与百合放入盘中，倒入花胶及汤汁。4. 最后放上鱼子酱即可。

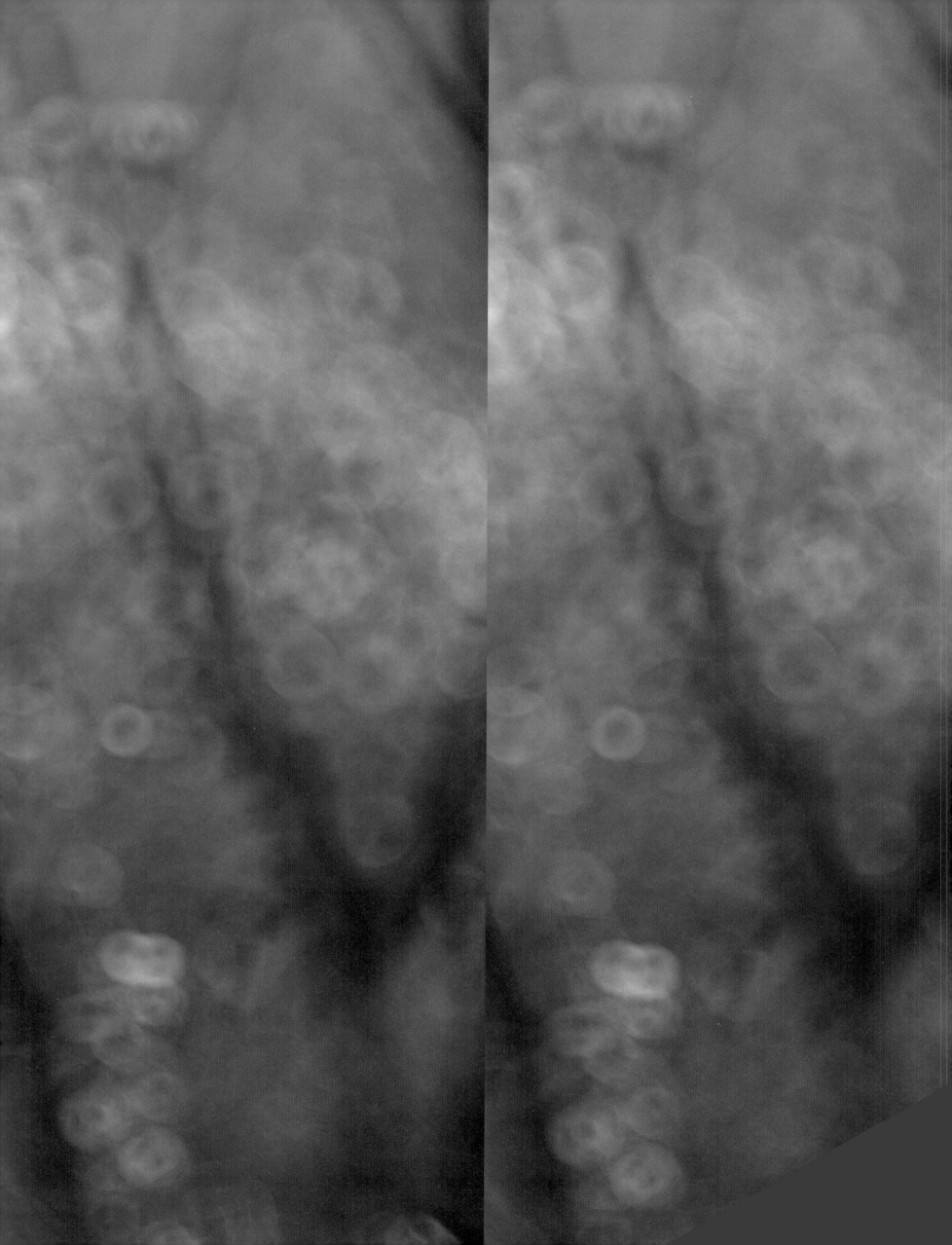

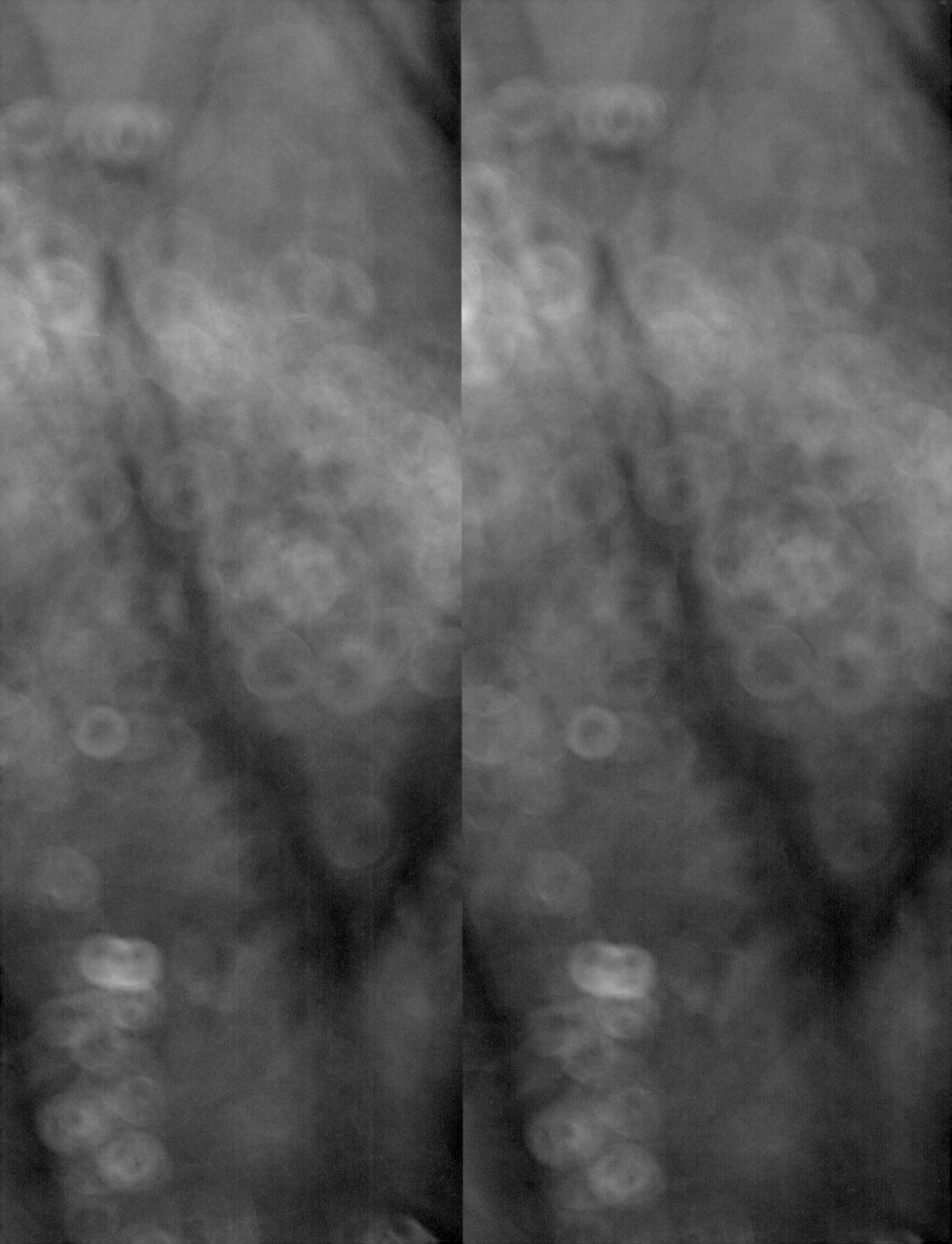

雪菜收获季节在霜降后，菜蔬经霜冻，水分析出，淀粉糖化，会甜香，像震泽的香青菜。雪菜用海盐腌，蛋白质转化为氨基酸，鲜类物质大量增多，炖汤、炒肉末，怎样做都好吃。大汤泡了雪菜，鲜上加鲜，好里又好。

——《大黄鱼蜕变》

Post-frost, the harvest of "snow vegetables" begins. These frost-kissed greens, akin to Zhenze town's "xiangqingcai" vegetable, transform their starches into sugars, becoming sweet and aromatic. Salted with sea salt, their proteins morph into umami-rich amino acids. Whether in soups or stir-fried with minced meat, their charm is undeniable. Adding the delightful touch of "snow vegetables" into a luscious broth elevates its allure to unparalleled heights.

—— **The Transformation of the Yellow Croaker**

霜降

Frost's Descent

THE 52RD PHENOLOGICAL PERIOD
Jackals make offerings of their prey

Jackals display their prey before consuming it.

52 / 72 物候
豺乃祭兽

豺狼把自己捕获而来的猎物先陈列好，之后再食用。

净洗青红　　骤飞沧海雨

宋 · 吴文英《齐天乐》

Braised Crab With Mung Bean Vermicelli
Preparation : 30 mins. Ingredients : crab 4pcs (chopped), mung bean vermicelli 200g (soaked), bird eye chili 6g (sliced). Seasonings : prawn bisque 500ml, salt 5g, corn starch 100g, sugar 7g, ground white pepper 3g, canola seed oil 20ml. Finishing : 30 mins. Method : 1. Coat crab with corn starch then deep fry in 180°C, until red color. 2. Fry chili with canola seed oil, add prawn bisque then season with ground white pepper, salt and sugar. 3. Add fried crab into the bisque to braise 5 mins, then add vermicelli braised for another 3mins.

青蟹烧龙口粉丝
准备时间：30 分钟。食材：青蟹 4 只，龙口粉丝 200 克（浸泡），朝天椒 6 克 (切细)。调料：虾汤 500 毫升，盐 5 克，玉米淀粉 100 克，白砂糖 7 克，白胡椒粉 3 克，菜籽油 20 毫升。制作时间：30 分钟。做法：1. 青蟹切成块，蘸裹玉米淀粉后以 180°C 热油炸至红色，备用。2. 锅中将朝天椒用菜籽油煸香，倒入虾汤后以白胡椒粉、盐、白砂糖调味。3. 将青蟹放入虾汤中烧 5 分钟，再放入龙口粉丝烧 3 分钟即可上盘。

Color Combination 色彩搭配：

犹余雪霜态　未肯十分红

宋·王十朋《红梅》

Alaska King Crab Rice

Preparation : 30 mins. Ingredients : Alaska king crab 1pc, black truffle 1pc, rice 600g, galangal 20g (finely chopped), ginger 10g (finely chopped), green garlic 10g (diced). Seasonings : chicken stock 500ml, chicken fat 25ml, steam fish soy sauce 30ml. Finishing : 40 mins. Method : 1. Steam Alaska king crab for 5mins, then peel the meat out of shell. 2. In clay pot, heat chicken oil and fry ginger, then add rice. 3. Add chicken stock, cover and cook in low heat around 15mins. 4. Spread galangal on rice, then arrange Alaska king crab meat on top, turn to high heat to make crust of cooked rice. 5. Garnish green garlic dices, dress steam fish soy sauce and shave black truffle while it serves.

帝王蟹焗饭

准备时间：30 分钟。食材：帝王蟹 1 只，黑松露 1 块，大米 600 克，沙姜蓉 20 克，姜米 10 克，青蒜粒 10 克。调料：鸡汤 500 毫升，鸡油 25 毫升，蒸鱼豉油 30 毫升。制作时间：40 分钟。做法：1. 帝王蟹隔水蒸 5 分钟至七成熟后，取出蟹肉备用。2. 砂锅中加热鸡油，放入姜米炒香后再放入大米，加入鸡汤，盖盖转小火将大米煮熟。3. 将沙姜蓉抹在米饭上，放上帝王蟹再转大火烧出锅巴。4. 食用时撒入青蒜粒，淋上蒸鱼豉油，刨上黑松露片即可。

Color Combination　色彩搭配：

THE 53RD PHENOLOGICAL PERIOD
Plants turn yellow and fall

53 / 72 物候
草木黄落

草木变黄枯落。

Vegetation yellows and sheds.

帐黄橘绿总寻常
看丹桂
余香再吐

宋·管鉴《鹊桥仙·八月二十八日寿唐子才》

Braised Porcini With Risoni

Preparation : 30 mins. Ingredients : porcini 50g, risoni 25g (soaked), brussels sprouts 10g (blanched), Parmasen cheese 30g (grinded), shallot 10g (finely chopped). Seasonings : abalone sauce 100ml, butter 20g, water 200ml, salt 5g, cream 50ml, chicken stock 100ml, extra virgin olive oil 5ml. Finishing : 30 mins. Method : 1. Toss brussels sprouts with extra virgin olive oil. 2. Pan fry porcini with 10g butter, add water and abalone sauce, braise for 5mins, keep warm. 3. Boil risoni for 5 mins, drain. 4. Sweat shallot with butter, add chicken stock and risoni, season with salt. 5. Add Parmasen cheese and cream, reduce until creamy, arrange on plate. 6. Top with porcini and brussels sprouts.

烧牛肝菌配意大利米面

准备时间：30 分钟。食材：牛肝菌 50 克，意大利米 25 克（泡发），抱子甘蓝 10 克，帕马森芝士 30 克（磨碎），红葱头 10 克（切碎）。调料：鲍汁 100 毫升，黄油 20 克，水 200 毫升，盐 5 克，淡奶油 50 毫升，鸡汤 100 毫升，特级初榨橄榄油 5 毫升。制作时间：30 分钟。做法：1. 抱子甘蓝与特级初榨橄榄油拌匀备用。2. 牛肝菌用 10 克黄油煎香，加入水、鲍汁，小火烧 5 分钟，保温备用。3. 意大利米放入沸水中煮 5 分钟后，沥干水分。4. 锅中将红葱头用黄油煸香，放入鸡汤和意大利米，并以盐调味。5. 加入淡奶油及帕马森芝士，待收汁后上盘。6. 放上牛肝菌及抱子甘蓝即可。

Color Combination　色彩搭配：

天边金掌露成霜 云随雁字长

宋·晏几道《阮郎归·天边金掌露成霜》

Braised Cabbage With Chestnut In Saffron Sauce
Preparation : 15 mins. Ingredients : baby Chinese cabbage 250g, chestnut 12 pcs. Seasonings : salt 2g, sugar 3g, saffron water 30ml (mix1g saffron with 30ml water), superior thick consommé 450ml. Finishing : 10 mins. Thickening Agent: corn starch 5g, water 10ml. Method : 1. Branch baby Chinese cabbage in boiling water, then shred by hand. 2. Boil the superior thick consommé, then mix with saffron water, season with salt and sugar. 3. Add the Chinese cabbage and chestnut, cook for 3 mins. 4. Add the thickening agent stir well, then serve.

红花汁栗子白菜
准备时间：15 分钟。食材：高山娃娃菜 250 克，板栗 12 颗。调料：盐 2 克，白砂糖 3 克，藏红花水 30 毫升（1 克藏红花：30 毫升纯净水），浓汤 450 毫升。制作时间：10 分钟。勾芡：淀粉 5 克，纯净水 10 毫升。做法：1．先将娃娃菜氽水，然后撕成条状备用。2．浓汤烧开，放入盐、白砂糖调味，再加入藏红花水调成金黄色。3．放入撕好的白菜与板栗，再烧制 3 分钟后勾芡即可。

Superior thick Consommé / Sanhuang chicken 1pc (1250g), ribs 250g, water 3.5l, put all ingredients into a pot, stew for 3 hours, then boil for another 30 mins, drain.

浓汤制作 / 将三黄鸡 1 只（约 1250 克）、排骨 250 克、纯净水 3.5 升一起以小火炖煮 3 小时，再转大火煮 30 分钟，过滤汤汁即成。

Color Combination 色彩搭配：

THE 54TH PHENOLOGICAL PERIOD
Hibernating insects go dormant

Insects enter their winter dormant state.

54 / 72 物候
蛰虫咸俯

■ 蛰伏的虫子进入冬眠蛰伏的状态。

人闲桂花落　　夜静春山空

唐 · 王维《鸟鸣涧》

Goose Liver With Caviar

Preparation : 1hr. Ingredients : goose liver 800g, caviar 20g. Seasonings : spice stock 2000ml. Finishing : 10 mins. Method : 1. Boil spice stock, then poach goose liver for 30mins, drain and cool. 2. Cut goose liver in cube, arrange on plate and garnish with caviar.

Spice Stock / hot water 2500ml. Ingredients: leek 200g, ginger 100g, garlic 100g, shallot 150g, coriander 50g, celery 80g. Spices: star anis 7g, galanga 2g, amomum tsaoko 2g, lemon grass 8g, fructus momordicae 1pc, bay leaf 7g, fennel seeds 6g, white pepper 4g, cinnamon 6g, Sichuan pepper 5g, dried orange peel 10g. Seasonings: salt 10g, sugar 5g, rock sugar 75g, dark soy sauce 30ml, light soy sauce 90ml, fish sauce 30ml, Shaoxing huadiao wine 20ml. Method: 1. Fry ingredients in a pot, add spices keep frying until fragrance. 2. Add seasonings and hot water, boil for 30mins ready to use.

汕头老鹅肝配鱼子酱

准备时间：1小时。食材：鹅肝 800 克，鱼子酱 20 克。调料：卤水 2000 毫升。制作时间：10 分钟。做法：1 将卤水烧开后关火，把鹅肝放入卤水中浸泡 30 分钟后沥干，摊凉。2 将鹅肝切成方块上盘，再放上鱼子酱即可。

卤水制作 / 热水 2500 毫升。食材：大葱 200 克，姜 100 克，大蒜 100 克，红葱头 150 克，香菜 50 克，芹菜 80 克。香料：八角 7 克，沙姜 2 克，草果 2 克，香茅草 8 克，罗汉果 1 个，香叶 7 克，小茴香 6 克，白胡椒 4 克，桂皮 6 克，花椒 5 克，陈皮 10 克。调料：盐 10 克，白砂糖 5 克，冰糖 75 克，老抽 30 毫升，生抽 90 毫升，鱼露 30 毫升，花雕酒 20 毫升。做法：1. 将食材放入锅中煸炒出香气，再加入香料炒制后加入调料。2. 加入热水，大火烧半小时后备用。

Color Combination　色彩搭配：

九月霜秋秋已尽 烘林败叶红相映

宋·欧阳修《渔家傲·九月霜秋秋已尽》

Color Of Autumn

Preparation : 3 hrs. Ingredients : tiramisu 100g, spicy sponge cake 100g, coco cookie crumble 100g, crispy maple leaf 30g, panna cotta 80g. Seasoning : icing sugar 5g. Finishing : 10 mins. Method : 1. Place tiramisu on place, side with spicy sponge cake, cover with coco cookie crumble. 2. Place crispy maple leaf and panna cotta. Finish with dusting icing sugar.

霜叶红于二月花

准备时间：3 小时。食材：提拉米苏 100 克，麻辣蛋糕 100 克，可可黑土曲奇 100 克，三色枫叶薄脆 30 克，奶油小脚冻 80 克。调料：糖粉 5 克。制作时间：10 分钟。做法：1. 先将提拉米苏放在盘中，边上点缀麻辣蛋糕，再覆盖上可可黑土曲奇碎。2. 将三色枫叶薄脆码放好后放上奶油小脚冻，再筛上糖粉即可。

Tiramisu / Ingredients : finger biscuit 300g, mascarpone 500g, cream 250ml. Seasonings : icing sugar 100g, expresso 500ml, rum 50ml, coco powder 50g. Method : 1. Stir mascarpone until smooth , then mix with cream and icing sugar. 2.Mix rum and expresso, then dip in finger biscuit. 3. Place finger biscuit in mould, layer mascarpone mixture, then keep in refrigerator for 2 hours. 4.Dust coco powder and slice into 2.5cmX2.5cm pieces.
Panna Cotta / Ingredients : cream 500ml, sugar 100g, vanilla stick 1pc, gelatin 3pcs (soaked), kirsch 20g. Method : 1.Heat cream, vanilla and sugar in a small pot, then add gelatin, stir to melt. 2. Add kirsch and stir well, pour into a feet mould, keep in refrigerator for 1 hour.
Spicy Sponge Cake / Ingredients : butter 25g, egg 2pcs, chocolate 100g (melted), sugar 100g, chili powder 20g, ground Sichuan pepper 15g, flour 180g. Method : 1.Beat butter, sugar and egg, slowly stir in chocolate, chili powder, ground Sichuan pepper and flour. 2. Pour the cake mixture into a baking tray, bake in 180°C for 15mins, keep cool. 3. Slice in pieces when it serves.
Coco Cookie Crumble / Ingredients : sugar 100g, almond powder 100g, flour 100g, coco powder 20g, salt 2g, butter 100g. Method : Mix all Ingredients, bake in 180°C for 8 mins, rub into crumble.
Crispy Maple Leaf / Ingredients :Valrhona hazelnuts praline 100g, butter 50g, egg 2pcs, flour 100g, icing sugar 50g, hawthorn purée 20g, passion fruit jam 20g, coffee powder 10g. Method :1.Beat Valrhona hazelnuts praline and butter, then mix with egg one by one. 2. Mix icing sugar and flour until smooth, separate into 3 portions. 3. Mix hawthorn purée, passion fruit jam and coffee powder with mixture to make different kinds of color. 4. Press the mixture into maple leaf mould, then place on a silicone baking map.5.Bake in 130°C for 10 mins, keep cool to serve.

提拉米苏 / 食材：手指饼干 300 克，马斯卡彭奶酪 500 克，奶油 250 毫升。调料：糖粉 100 克，意式特浓咖啡 500 毫升，朗姆酒 50 毫升，可可粉 50 克。做法：1. 马斯卡彭奶酪搅拌顺滑，慢慢拌入糖粉及奶油，备用。2. 将朗姆酒与意式特浓咖啡混合。3. 将手指饼干蘸入咖啡液，放入模具，再铺上马斯卡彭奶酪，再放上手指饼干，再铺上马斯卡彭奶酪，放入冰箱冷藏 2 小时。4. 将可可粉撒在上面，再切成 2.5×2.5 厘米的方块。
奶油小脚冻 / 食材：淡奶油 500 毫升，白砂糖 100 克，香草荚 1 根，鱼胶片 3 片（泡软），樱桃白兰地 20 毫升。做法：1. 将淡奶油与白砂糖、香草荚放到小锅中小火烧开，放入 3 片鱼胶片，放凉。2. 缓慢倒入樱桃白兰地并搅拌均匀，再灌入小脚型的模具中。3. 放入冰箱冷冻 1 小时即可。
麻辣蛋糕 / 食材：黄油 25 克，鸡蛋 2 个，巧克力 100 克（融化），白砂糖 100 克，辣椒粉 20 克，花椒粉 15 克，面粉 180 克。做法：1. 黄油、白砂糖和鸡蛋打发，然后慢慢拌入巧克力、辣椒粉、花椒粉、面粉。2. 将拌好的蛋糕浆倒入烤盘，以 180°C温度烤 15 分钟，晾凉。3. 食用时切成方粒。
可可黑土曲奇 / 食材：白砂糖 100 克，杏仁粉 100 克，面粉 100 克，可可粉 20 克，盐 2 克，黄油 100 克。做法：将所有食材混合，放入烤箱180°C烤 8 分钟，取出搓捏成粉即可。
三色枫叶 / 食材：法芙娜榛果酱 100 克，黄油 50 克，鸡蛋 2 个，面粉 100 克，糖粉 50 克，山楂蓉 20 克，热情果酱 20 克，咖啡粉 10 克。做法：1. 将法芙娜榛果酱和黄油打发，分次加入鸡蛋拌匀后再加入面粉和糖粉拌匀。2. 将面团分成 3 份，分别混合山楂蓉、热情果酱及咖啡粉。3. 将三种颜色的面团填压到枫叶形状的模具中，码放在硅胶垫上。4. 将枫叶放入烤箱 130°C烤 10 分钟即成。

Color Combination 色彩搭配：

糖炒栗子的质感和味感，用一『甘』字，足矣。它没有那么甜，有一种君子恬淡之风。质感上也有『肥甘厚实』的咬劲，带着植物果仁特有的香气。

——《秋冬至景，板栗『甘』味》

The allure of candied chestnuts lies in their sweetness, a subtlety that whispers rather than shouts. They are satisfyingly substantial in texture and infused with the earthy aroma of nuts.

—**Autumn and Winter Delight: The 'Sweet' Taste of Chestnuts**

立冬

Beginning

of

Winter

THE 55TH PHENOLOGICAL PERIOD
Water starts to freeze

Ice begins to form on water surfaces.

55 / 72 物候
水始冰

水开始凝结成冰。

皑如山上雪　皎若云间月

汉 · 卓文君《白头吟》

Crispy Cabbage
Preparation : 10 mins. Ingredients : Chinese cabbage 200g. Seasonings : rice vinegar 10ml, sugar 5g, sesame oil 1ml, garlic 1g (finely chopped). Finishing : 20 mins. Method : 1. Wash Chinese cabbage, slice in thin, soak into the ice water. 2. Mix the seasonings into dressing, set aside. 3. Drain the Chinese cabbage slice, toss with the dressing.

凤尾白菜
准备时间：10 分钟。食材：白菜 200 克。调料：米醋 10 毫升，白砂糖 5 克，香油 1 毫升，蒜蓉 1 克。制作时间：20 分钟。做法：1. 先将白菜洗净，改花刀后放入冰水中浸泡。2. 把调料混合调成糖醋汁，备用。3. 将白菜捞出，控干水分后与糖醋汁拌匀即可。

Color Combination　色彩搭配：

黑潭水深黑如墨　　传有神龙人不识

　　　　　　　　　唐 · 白居易《黑潭龙 · 疾贪吏也》

Tofu Julienne In Squid Ink Soup

Preparation : 10 mins. Ingredients : lactone bean curd 1 pc, asparagus lettuce 200g. Seasonings : salt 1g, chicken stock 300ml, squid ink 0.5ml, pickle juice 100ml, salt water 1l (1:10). Finishing : 10 mins. Method : 1. Slice the asparagus lettuce in thin julienne, soak in pickle juice. 2. Slice the lactone bean curd in thin julienne, soak in salt water. 3. Heat the chicken stock and stir in the squid ink, season with salt. 4. Add all the juliennes, stir carefully 2 mins in a low heat. 5. Pour into a soup plate and ready to serve.

墨鱼汁文思豆腐

准备时间：10分钟。食材：嫩豆腐1盒，莴笋200克。调料：盐1克，鸡汤300毫升，墨鱼汁0.5毫升，泡菜汤100毫升，盐水1升（1:10）。制作时间：10分钟。做法：1. 莴笋切细丝，用泡菜汤浸泡，备用。2. 嫩豆腐切丝，浸入盐水中备用。3. 鸡汤烧开后加盐调味，加入墨鱼汁搅匀。4. 最后加入豆腐丝和莴笋丝，以慢火煮2分钟即可。

THE 56TH PHENOLOGICAL PERIOD
The ground starts to freeze

The earth begins to harden from the cold.

56 / 72 物候
地始冻

土地也开始冻结。

疏雨洗天清
枕簟凉生
井桐一叶做秋声

宋·邓剡《浪淘沙·疏雨洗天清》

Hawthorn Foie Gras

Preparation : 15 mins. Ingredients : pate foie gras 200g, beet root juice 300ml, hawthorn 30g. Seasonings : sugar 130g, raspberry vinegar 100ml, gelatin sheet 15g. Finishing : 60 mins. Method : 1. Place the pate foie gras in a half ball shape mould, freeze for 40 mins, then form into a ball shape. 2. Pour the beet root juice and remaining seasonings in a small sauce pot, heat in low fire. 3. When the gelatin sheet gets dissolve, keep warm. 4. Dip the hawthorn and pate foie gras into the beet root juice, and set aside until the skin gets set.

山楂鹅肝

准备时间：15 分钟。食材：鹅肝酱 200 克，甜菜头汁 300 毫升，山楂 30 克。调料：白砂糖 130 克，树莓醋 100 毫升，鱼胶片 15 克。制作时间：60 分钟。做法：1．将鹅肝酱放入半圆形的模具中速冻 40 分钟，组合成球状。备用。2．把甜菜头汁与调料混合，倒入小锅中，以小火烧开，保温。3．将山楂及鹅肝酱放入步骤 2 中蘸匀，放在阴凉处凝固即可。

Color Combination 色彩搭配：

一别如斯　落尽梨花月又西

清 · 纳兰性德《采桑子 · 当时错》

White Chocolate Sea Shell

Preparation : 20 mins. Ingredients : white chocolate 250g (melt in double boiler), philadelphia cream cheese cream 250g, cream 250g, sugar 50g, honey 80ml, gelatin sheet 1pc (soaked). Seasonings : **spicy**: ground green Sichuan pepper 5g, ground chili powder 5g, ground Sichuan pepper 5g. **orange**: orange juice 20ml, corn starch 2g. **nutmeg**: ground nutmeg 10g, red bean paste 4g. **dried peel**: dried orange peel 25g (steamed and mashed). **coffee**: ground coffee 5g, cream 10g. Finishing : 150 mins. Method :1. Mix all the seasonings for different flavors of chocolate. 2. Fill the melted white chocolate in sea shell mould, then add different flavors mixture. Wait for set. 3. Heat Philadelphia cream cheese, cream, sugar, honey and gelatin sheet, wait for set. 4. Demold white chocolate sea shell, then arrange together with cheese cake mixture.

贝壳邂逅白芝士巧克力

准备时间: 20 分钟。食材: 白巧克力 250 克（隔水融化），卡夫菲力奶油芝士 250 克，淡奶油 250 克，白砂糖 50 克，蜂蜜 80 毫升，鱼胶片 1 片。调料: **麻辣味**: 麻椒粉 5 克，辣椒粉 5 克，花椒粉 5 克；**橙子**: 橙汁 20 毫升，玉米淀粉 2 克；**豆蔻味**: 肉蔻粉 10 克，红豆沙 4 克；**陈皮味**: 陈皮泥 25 克；**咖啡味**: 咖啡粉 5 克，奶油 10 克。制作时间: 150 分钟。做法: 1. 先将调料各自混合，加热，备用。2. 将白巧克力灌入贝壳形状的模具，然后分别加入各种口味。再灌入白巧克力封口，待冷却凝固。3. 将卡夫菲力奶油芝士与加热后的淡奶油、蜂蜜、白砂糖以及鱼胶片混合，待凝固后即成奶酪蛋糕。4. 将凝固的白巧克力贝壳上盘，再配上奶酪蛋糕即可。

Color Combination　色彩搭配:

57 / 72 物候
雉入大水为蜃

THE 57TH PHENOLOGICAL PERIOD
Pheasants turn into giant clams

古人认为雉到立冬后就变成了大蛤。

Ancient people believed that, come winter, pheasants transformed into giant clams

江醪白蚁醇
海馔糖蟹肥

宋·黄庭坚《次韵师厚食蟹》

Crab Pudding

Preparation : 5 mins. Ingredients : cooked drunken crab 1pc (peel meat and roe), baguette 50g. Seasonings : Champagne jelly 100g, molecular passion fruit caviar 30g. Finishing : 60 mins. Method : 1. Place crab meat and crab roe in mould and freeze for 50 minutes. Dress Champagne jelly, set. 2. Toast baguette and break into crumbs. 3. Combine frozen crab paste and bread crumbs, plate and top with passion fruit caviar.

蟹糊布丁

准备时间：5 分钟。食材：熟醉蟹 1 只（取蟹肉及蟹黄），法棍面包 50 克。调料：香槟果冻 100 克，热情果鱼子 30 克。制作时间：60 分钟。做法：1. 将蟹肉和蟹黄装入模具，放入冰箱冷冻 50 分钟。点缀上香槟果冻，备用。2. 法棍面包烤酥，掰碎。3. 将冻好的蟹糊和面包碎一起装盘，撒上热情果鱼子即可。

Color Combination　色彩搭配：

摇扇对酒楼　持袂把蟹螯

唐 · 李白《送当涂赵少府赴长芦》

Steamed Alaska King Crab With Meat Cake

Preparation : 30 mins. Ingredients : Alaska king crab 1pc (2.5kg), pork belly 250g (finely chopped), Sichuan bacon 75g (small diced), green soy bean (roughly chopped), egg white 25g. Seasonings : chicken stock 100ml, chicken fat 5ml, soy sauce for seafood 20ml, salt 2g, ground white pepper 2g, leek and ginger water 120ml. Finishing : 30 mins. Method: 1.Mix pork belly, Sichuan bacon, salt, ground white pepper, leek and ginger water, egg white until the texture become firm, keep in refrigerator. 2.Mix green soy bean and meat mixture, pan-fry until golden color, move to a plate. 3.Steam Alaska king crab for 5mins, peel the meat, then toss with soy sauce for seafood, then steam together with meat cake for 3mins. 4.Boil chicken stock and chicken fat, then dress on the crab while it serves.

青豆肉饼帝王蟹

准备时间：30 分钟。食材：阿拉斯加蟹一只（约 2.5 公斤），五花肉 250 克（剁馅），四川腊肉 75 克（切细粒），毛豆仁 50 克（切粗粒），蛋清 25 克。调料：鸡汤 100 毫升，鸡油 5 毫升，蒸鱼豉油 20 毫升，盐 2 克，白胡椒粉 2 克，葱姜水 120 毫升。制作时间：30 分钟。做法：1. 将五花肉馅、四川腊肉粒、盐、白胡椒粉、葱姜水、蛋清混合，顺着一个方向搅拌上劲，放入冰箱备用。2. 将肉馅与毛豆仁搅拌均匀，用煎锅煎出焦面，放入盘中。3. 阿拉斯加蟹整只蒸 5 分钟，取出蟹肉并用蒸鱼豉油拌匀，与肉饼一起放入蒸箱中蒸 3 分钟。4. 最后将鸡汤和鸡油烧开，淋在蒸好的蟹肉上即可。

Color Combination　色彩搭配：

苏式「酱方」到了北京，早已面目全非。想得开的，合自己口味儿就好；想不开的，就在正宗上较劲儿。

——《苏灶肉（苏造肉）的堕落》

The Suzhou-style "braised pork belly" has evolved since its arrival in Beijing, prompting a culinary debate. While some embrace the adaptation to local palates, purists lament the loss of authenticity.

— **The Culinary Evolution of Suzhou Zhaorou (Suzhou-style Braised Meat)**

小雪

Lesser Snow

THE 58TH PHENOLOGICAL PERIOD
Rainbows vanish

Rainbows no longer appear.

58 / 72 物候
虹藏不见

彩虹不再出现。

孤 舟 蓑 笠 翁　　独 钓 寒 江 雪

唐 · 柳宗元《江雪》

Sweet Pork Ribs At Snowy River
Preparation : 10 mins. Ingredients : pork rib 500g, preserved plum (huamei) 5g. Seasonings : salt 2g, chicken stock 300ml, dark soy sauce 10ml, sugar 30g, rice vinegar 50ml, star anise 2g, vegetable oil 5ml, icing sugar 5g. Finishing : 45 mins. Method : 1. Chopped the pork ribs in small piece, blanch in hot water. Drain. 2. Fry sugar and oil in a wok, until it starts to caramelize. 3. Add pork ribs and the remaining seasonings (except icing sugar), then simmering for 30 mins, until the sauce becomes syrup. Move on the plate. 4. Dust the icing sugar on the pork ribs when it serves.

江雪糖醋小排
准备时间：10 分钟。食材：猪小排 500 克, 话梅 5 克。调料：盐 2 克，鸡汤 300 毫升，老抽 10 毫升，白砂糖 30 克，米醋 50 毫升，大料 2 克，色拉油 5 毫升, 糖霜 5 克。制作时间：45 分钟。做法：1. 将猪小排剁成小块，汆水备用。2. 锅中用色拉油和白砂糖炒糖色，然后放入步骤 1 翻炒上色。3. 放入除糖霜以外剩余的调料和话梅，再以小火烧制 30 分钟，直至酱汁收浓即可。4. 食用时撒上糖霜即可。

Color Combination　色彩搭配：

316

莲 花 兜 上 草 虫 鸣　　处 处 村 庄 白 菜 生

明 · 郑明选《沈长山山庄绝句三首》

Pickled Cabbage In Dual Flavors

Preparation : 15 mins. Ingredients : Chinese cabbage 1500g, crab roe 3g, fried chili 1g. Seasonings : sesame oil 5ml, salt 2g, white vinegar 4ml, sugar 5g. Marinated Sauce 1: soy sauce 20ml, sugar 150g, rice vinegar 135ml, salt 5g, chili oil 5ml. Marinated Sauce 2: yellow mustard powder 25g, pure water 50ml. Finishing :13 hrs. Method : 1. Wash Chinese cabbage, cut in two, then blanch in boiling water, drain. 2. Squeeze the remaining water from the cabbage, then marinate half part with marinated sauce 1, keep in refrigerator for 12 hours. 3. Boil pure water, mix well with yellow mustard powder, keep in room temperature for 12 hours. 4. Mix remaining seasonings with mustard mixture, then marinate with another half cabbage. 5. Cut the sweet & sour cabbage and mustard cabbage in small chunk, put on the plate. 6. Garnish the crab roe on the mustard cabbage, and the fried chili on the sweet & sour cabbage.

白菜双墩

准备时间：15 分钟。食材：大白菜 1500 克，蟹子 3 克，油炸辣椒 1 克。调料：香油 5 毫升，盐 2 克，白醋 4 毫升，白砂糖 5 克。腌料 1：酱油 20 毫升，白砂糖 150 克，米醋 135 毫升，盐 5 克，干辣椒油 5 毫升。腌料 2：黄芥末粉 25 克，纯净水 50 毫升。制作时间：13 小时。做法：1. 大白菜洗净，从根部切开一分为二，氽水备用。2. 捏干步骤 1 的水分，一半与腌料 1 拌匀，放入冰箱腌制 12 小时。3. 将腌料 2 中的黄芥末粉用烧开的纯净水混合，常温放置 12 小时。4. 把调料与步骤 3 混合，再和另一半白菜混合，做成芥末白菜墩。5. 把做好的糖醋白菜墩和芥末白菜墩用刀切成段。6. 将步骤 5 上盘，把蟹子放在芥末白菜墩上，油炸辣椒放在糖醋白菜墩上。

Color　色彩
Combination　搭配：

THE 59TH
PHENOLOGICAL PERIOD
The heavens ascend

The connection between heaven and earth weakens, and all living things lose their vitality.

59 / 72 物候
天气上升

天地不通、阴阳不交,万物失去生机。

Sichuan Spicy Pancake Soup

Preparation : 10 mins. Ingredients : flour 200g, XO Sauce 50g, ground lemon grass 5g, soup stock 2000ml. Seasonings : salt 10g, ground paprika 10g, mince garlic 50g. Finishing : 1 hr. Method : 1. Mix water with flour into dough, slice in strips, form into pan cake then fry until both side are golden color. 2. Saute mince garlic and XO sauce, add remaining ingredients to make spicy soup, serve with fried pancake.

川辣皇转转饼汤

准备时间:10 分钟。食材:面粉 200 克,XO 酱 50 克,香茅粉 5 克,高汤 2000 毫升。调料:盐 10 克,甜椒粉 10 克,蒜蓉 50 克。制作时间:1 小时。做法:1. 面粉加水制成面条,盘成饼状,煎至两面金黄。2. 起锅焗蒜蓉、XO 酱,加入高汤、盐、香茅粉、甜椒粉制成汤,与煎饼摆盘。

新粟米炊鱼子饭
嫩芦笋煮鳖裙羹

清・李渔《闲情偶记》

Color Combination 色彩搭配:

酸甜红颗阿谁知　别是人间滋味

宋·周紫芝《西江月·席上赋双荔子》

Rose Radish

Preparation : 10 mins. Ingredients : radish 500g. Seasonings : Sichuan pickle juice 300ml, beetroot juice 100ml, salt 10g. Finishing : 50 mins. Method : 1. Slice half of radish into thin and flat, and the other half into thin julienne. 2. Mix the salt with radish then set for 10 mins, gently squeeze the water of the radish. 3. Mix Sichuan pickle juice and beetroot juice, then soak the radish for 30mins. 4. Squeeze extra juice of the radish, then roll the julienne into the flat radish. 5. Slice the radish roll while it serves.

泡玫瑰萝卜

准备时间：10 分钟。食材：白萝卜 500 克。调料：四川泡菜水 300 毫升，红菜头汁 100 毫升，盐 10 克。制作时间：50 分钟。做法：1. 白萝卜分别切成细丝与大片，用盐拌匀，腌渍 10 分钟。2. 将腌渍好的萝卜捏出多余的水分，备用。3. 将四川泡菜水兑入红菜头汁，放入两种萝卜分别浸泡约半小时。4. 将泡好的萝卜丝放入萝卜片中卷成卷，斜刀切成块，装盘即可。

Color Combination　色彩搭配：

THE 60TH PHENOLOGICAL PERIOD
Winter's grip tightens
The world enters the deep cold of winter

60 / 72 物候
闭塞而成冬

天地闭塞，转入严寒的冬天。

湖堤疏瘦水杨柳　　村舍殷红山石榴

宋·陆游《杂题》

Highland Barley Pomegranate Salad

Preparation : 10 mins. Ingredients : highland barley 10g, pomegranate seed 30g, mix salad 100g. Seasonings : salt 4g, sugar 3g, aged vinegar 5ml, extra virgin olive oil 10ml. Finishing : 40 mins. Method : 1. Steam highland barley with water for 30mins and let cool. 2. Mix seasonings into vinaigrette. 3. Toss steamed highland barley, pomegranate seeds and mix salad with vinaigrette when serve.

青稞石榴沙拉

准备时间：10分钟。食材：青稞10克，石榴籽30克，混合生菜100克。调料：盐4克，白砂糖3克，陈醋5毫升，特级初榨橄榄油10毫升。制作时间：40分钟。做法：1.青稞放入少许水，蒸制30分钟后摊凉备用。2.将盐、白砂糖、陈醋与橄榄油混合成汁，备用。3.把青稞、石榴籽及混合生菜与步骤2拌匀，装盘即可。

Color Combination　色彩搭配：

随缘柳绿柳白
费尽雕镂
疏林野水
任横斜
谁与妆修

宋·陈亮《汉宫春·雪月相投》

Pickled Wild Kohlrabi Radish

Preparation : 5 mins. Ingredients :1 kohlrabi radish 1pc. Seasonings : mustard paste 5g, white vinegar 5ml, Sichuan pepper 1g, salt 1g. Finishing : 20 mins. Method : 1. Shred kohlrabi radish. 2. Mix mustard, Sichuan pepper, salt, and white vinegar, marinate radish in mixture for 15 minutes. 3. Form the marinated radish into balls and plate as shown.

憋辣菜

准备时间：5 分钟。食材：憋辣菜 1 棵。调料：芥末膏 5 克，白醋 5 毫升，四川花椒 1 克，盐 1 克。制作时间：20 分钟。做法：1. 憋辣菜刨丝备用。2. 将白醋、芥末膏、四川花椒、盐调成汁，放入憋辣菜腌 15 分钟。3. 将腌制好的憋辣菜上盘即可。

猪腰切得越多，心念越成，日积月累，枯燥乏味成乐趣横生，刀工犹如神助，心念即成神明。刀就是神明，神明即刀。

——《怂厨子做不出爆炒腰花》

The art of slicing pig kidneys is a study of patience and precision. With time, the repetitive motion transforms into a meditative ritual, your knife skills approaching a zen-like mastery. The knife becomes an extension of your intent, a conduit of culinary creativity.

—— **A Timid Chef Can't Make Stir-Fried Kidney Slices**

大雪

Greater

Snow

THE 61ST PHENOLOGICAL PERIOD
Flying squirrels fall silent

Flying squirrels cease to sing.

61 / 72 物候
鹖鴠不鸣

■ 寒号鸟不再鸣叫。

云日相辉映　　空水共澄鲜

南北朝 · 谢灵运《登江中孤屿》

Braised Conch And Porcini With Truffle And Risoni

Preparation : 3 days. Ingredients : dried conch 50g (soaked for 3 days), porcini 50g, risoni 50g (cooked), tomato heart 5g (peeled in 1cm ball), shallot 5g (finely chopped). Seasonings : truffle sauce 10g, butter 15g, chicken soup stock 2.1l, oyster sauce 50ml, rock sugar 10g, dark soy sauce 10ml, Maggi 2ml, Parmasen cheese 5g (grinded), cream 10g, milk 10ml, salt 2g. Finishing : 3 hrs. Method : 1. Cut porcini in half, pan-fry with 10g of butter, keep aside. 2. Steam conch with 2l of chicken soup stock, oyster sauce, rock sugar, dark soy sauce and Maggi for 2hrs. 3. Drain steamed conch, then braise together pan-fry porcini with truffle sauce and 300ml of steam conch juice in low heat for 10mins, until it becomes syrup texture. 4. Heat risoni with remaining butter, shallot and remaining chicken soup stock. 5. Add salt, Parmasen cheese, cream and milk, heat until it becomes thick. Then arrange on plate. 6. Arrange braised conch and porcini on top of the risoni, garnish with tomato heart to serve.

黑松露酱烧大响螺黄牛肝菌配意大利米面

准备时间：3 天。食材：干大响螺半个（50 克，泡 3 天），黄牛肝菌 50 克，意大利米形面 50 克（煮熟），番茄芯球 5 克（直径 1 厘米左右），红葱 5 克（切碎）。调料：松露酱 10 克，黄油 15 克，鸡汤 2.1 升，蚝油 50 毫升，冰糖 10 克，老抽 10 毫升，美极鲜 2 毫升，帕尔马奶酪碎 5 克，淡奶油 10 克，牛奶 10 毫升，盐 2 克。制作时间：3 小时。做法：1. 黄牛肝菌切半，用 10 克黄油煎香，备用。2. 将泡发的大响螺，与鸡汤 2 升、蚝油、冰糖、老抽以及美极鲜放在碗中蒸 2 小时。3. 取出海螺，与煎好的牛肝菌以及松露酱、300 毫升的蒸海螺汁一起以小火烧 10 分钟，收汁。4. 意大利米形面与剩余的黄油、红葱、剩余的鸡汤一起烧开。5. 加入盐、帕尔马奶酪碎、淡奶油及牛奶，烧至黏稠，装盘。6. 把海螺与牛肝菌放在意大利米形面上，伴番茄心球即可。

Color Combination　色彩搭配：

落日水熔金　天淡暮烟凝碧

宋 · 廖世美《好事近 · 夕景》

Egg

Preparation : 5 mins. Ingredients : gelatine sheet 2pcs, xylitol 20g, egg white 50g, frozen persimmon heart 20g. Finishing : 1 hr. Method : 1. Mix gelatine sheet, xylitol and egg white to make egg shell. Let set for 45 minutes. 2. Place frozen persimmon heart into egg shell.

摔碎了的蛋

准备时间：5 分钟。食材：鱼胶片 2 片，木糖醇粉 20 克，鸡蛋清 50 克，冻柿子心 20 克。制作时间：1 小时。做法：1. 将鱼胶片、木糖醇粉、鸡蛋清混合，制成蛋壳状，静置 45 分钟。2. 将冻柿子心放入蛋壳中，装盘即可。

Color Combination　色彩搭配：

THE 62ND PHENOLOGICAL PERIOD
Tigers begin to mate

Tigers start their mating season.

62 / 72 物候
虎始交

老虎开始求偶。

东篱把酒黄昏后　有暗香盈袖

宋 · 李清照《醉花阴 · 薄雾浓云愁永昼》

Crispy Meat Pancake With Cheese

Preparation : 15 mins. Ingredients : pork belly 200g, cheddar cheese slice 100g, flour 250g. Seasonings : dark soy sauce 3ml, light soy sauce 20ml, salt 1g, five spices powder 0.5g, chopped shallots 30g, chopped ginger 10g, ground pepper 0.5g. Finishing : 60 mins. Method : 1. Finely dice pork belly and mix with dark and light soy sauces, salt, five spices powder, ginger, shallots, and ground pepper to make stuffings. 2. Mix flour and water to make dough and form into pancakes, fill with cheddar cheese and meat stuffings. 3. Bake pancakes at 220°C for 15 minutes.

书皮肉饼

准备时间：15 分钟。食材：五花肉 200 克，车打芝士片 100 克，面粉 250 克。调料：老抽 3 毫升，酱油 20 毫升，盐 1 克，五香粉 0.5 克，葱末 30 克，姜末 10 克，胡椒粉 0.5 克。制作时间：60 分钟。做法：1. 五花肉切小丁，加入所有调料腌好，搅打成肉馅备用。2. 面粉加水和软，加入肉馅及车打芝士片制成饼坯。3. 电饼铛上下温度调至 220℃，放入肉饼烙 15 分钟即可。

Color Combination 色彩搭配：

夜深未觉清香绝　　风露落溶月

宋 · 范成大《虞美人 · 红木犀》

Steam Hairy Crab Roe With Egg

Preparation : 15 mins. Ingredients : hairy crab roe (female crab) 20g, hairy crab white roe (male crab) 20g, egg 2pcs, shrimp soup 50ml, molecular balsamic caviar 3g. Seasonings : salt 5g, ginger juice 10ml. Finishing : 30 mins. Method : 1. Mix egg, shrimp soup and salt, steam for 15minutes to cook. 2. Saute crab roe with ginger juice. 3. Place the crab roe, molecular balsamic caviar on steam egg.

蛋羹秃黄油

准备时间：15分钟。食材：蟹黄（母蟹）20克，蟹膏（公蟹）20克，鸡蛋2个，虾汤50毫升，黑醋鱼子3克。调料：盐5克，姜汁10毫升。制作时间：30分钟。做法：1. 鸡蛋打碎，加入虾汤、盐，蒸15分钟制成蛋羹。2. 在蟹黄、蟹膏中加入姜汁，炒成秃黄油。3. 在蛋羹上撒上秃黄油、黑醋鱼子即可。

Color Combination　色彩搭配：

新雨后 天气晚来秋
松间照 清泉石上流
归浣女 莲动下渔
春芳歇 王孙自可

唐·王

THE 63RD PHENOLOGICAL PERIOD
Orchids sprout

Orchids, sensing the Yang energy, begin to grow.

63 / 72 物候
荔挺出

兰草感到阳气萌动而抽出新芽。

Gentleman's Soup
Preparation : 15 mins. Ingredients : xo's penis 50g, soft shell turtle 30g, hen meat 30g. Chinese herbals: goji berry 0.5g, American ginseng 0.5g, morinda 1g, eucommia ulmoides 0.5g, ophiopogon root 1g, codonopsis pilosula 0.5g, cartialgenous 0.2g(2-3 slices). Seasonings : salt 2g, chicken soup stock 220ml. Finishing : 4 hrs 30 mins. Method : 1. Blanch all Ingredients and Chinese herbals, drain. 2. Add soft shell turtle, hen meat and Chinese herbals together with chicken stock into a steam soup bowl. Then steam for 3 hours. 3. Add xo's penis into steam soup bowl, then steam for another 1 hour. 4. Season with salt while it serves.

男人汤
准备时间：15 分钟。食材：牛鞭 50 克，甲鱼 30 克，老鸡肉 30 克。药料：干枸杞 0.5 克，洋参片 0.5 克，巴戟 1 克，杜仲 0.5 克，麦冬 1 克，党参 0.5 克，鹿茸 0.2 克（2~3 片）。调料：盐 2 克，鸡汤 220 毫升。制作时间：4 小时 30 分钟。做法：1. 将甲鱼、老鸡肉、牛鞭及药料汆水，沥干备用。2. 将甲鱼、老鸡肉及药料与鸡汤倒入汤罐中，放入蒸箱中蒸 3 小时。3. 将汤罐取出，放入牛鞭再蒸 1 小时。4. 用盐调味即可。

丰神飘洒
器宇轩昂

明·罗贯中《三国演义》

Color Combination 色彩搭配：

芙蓉不及美人妆
水殿风来珠翠香

唐·王昌龄《西宫秋怨》

Lady's soup

Preparation :15 mins. Ingredients : soft shell turtle 40g, hen meat 30g, soft shell turtle skirt edge 40g, black chicken 60g, tendon 20g. Chinese herbals : American ginseng 1g, goji berry 2g, codonopsis pilosula 1g, dried longan 2g. Seasonings: chicken stock 220ml, salt 1g. Finishing : 4 hrs 30 mins. Method :1. Blanch all Ingredients and Chinese herbals, drain. 2. Add soft shell turtle, hen meat, black chicken and Chinese herbals together with chicken stock into a steam soup bowl. Then steam for 3 hours. 3. Add soft shell turtle skirt edge and tendon into steam soup bowl, then steam for another 1 hour. 4. Season with salt while it serves.

女人汤

准备时间：15 分钟。食材：甲鱼 40 克，老鸡肉 30 克，裙边 40 克，乌鸡 60 克，蹄筋 20 克。药料：洋参片 1 克，干枸杞 2 克，党参 1 克，桂圆肉 2 克。调料：清鸡汤 220 毫升，盐 1 克。制作时间：4 小时 30 分钟。做法：1. 将甲鱼、老鸡肉、乌鸡及药料氽水，沥干备用。2. 将甲鱼、老鸡肉、乌鸡、药料与清鸡汤放入汤罐，放入蒸箱中蒸 3 小时。3. 将汤罐取出，放入裙边及蹄筋，再蒸 1 小时。4. 用盐调味即可。

大白菜有松树的风格，不畏严寒，经过霜打，甘甜味美。经过老百姓的巧手，调出胜味。大白菜的味道是平常，平常心是道，有了平常心，日子能过得有滋有味。

——《大白菜的味道是平常，平常心就是道》

The humble Chinese cabbage, much like a steadfast pine tree, remains defiant in the face of bitter cold. The frost only serves to enhance its sweetness, teasing out a deeper, more complex flavor. The common folk, with their seasoned hands and innate culinary wisdom, coax out the cabbage's best attributes. The taste may be simple, but therein lies the beauty. Embracing simplicity is the path to contentment.

— **The Unassuming Elegance of Chinese Cabbage: A Lesson in Culinary Simplicity**

冬至

Winter

Solstice

347

THE 64TH PHENOLOGICAL PERIOD
Earthworms form knots

64 / 72 物候
蚯蚓结

As Yang energy starts to rise, earthworms turn upwards, their bodies resembling knotted ropes.

阳气微生，蚯蚓的头开始转而向上，身体的形状像是打了结的绳子。

长江绕郭知鱼美
好竹连山觉笋香

宋·苏轼《初到黄州》

Sour and Spicy Mandarin Fish

Preparation : 10 mins. Ingredients : mandarin fish fillet 750g, egg white 1pc, spring onion 15g (finely chopped), coriander 15g (finely chopped) . Seasonings : salt 3g, ground white pepper 8g, corn starch 5g, meat stock 2500ml, sesame oil 10ml, rice vinegar 15ml. Finishing : 20 mins. Method : 1. Slice mandarin fish fillet into flat thin slices, mix with salt, ground white pepper, corn starch and egg white. 2. Boil meat stock, add rice vinegar and sesame oil. 3. Add mandarin fish slices, serve with spring onion and coriander.

明炉醋椒鳜鱼

准备时间：10 分钟。食材：鳜鱼肉 750 克，蛋清 1 个，小葱末 15 克，香菜末 15 克。调料：盐 3 克，白胡椒粉 8 克，玉米淀粉 5 克，高汤 2500 毫升，芝麻油 10 毫升，米醋 15 毫升。制作时间：20 分钟。做法：1. 鳜鱼带皮片成大薄片，与盐、白胡椒粉、玉米淀粉及蛋清拌匀，备用。2. 高汤烧开，加入米醋及芝麻油。3. 放入鳜鱼片，氽熟后装入碗中，配小葱末及香菜末即可。

Color Combination 色彩搭配：

波心荡　冷月无声

宋 · 姜夔《扬州慢 · 淮左名都》

Double Boiled Shark Fin With Almond Cream

Preparation : 10 mins. Ingredients : almond 100g, shark fin 150g. Seasonings : salt 4g, sugar 6g, ground white pepper 2g, milk 20ml, pure water 300ml, glutinous rice flour 3g, corn starch 2g. Finishing:10 hrs. Method : 1. Soak almond and shark fin into water for 8 hours. 2. Wash Shark Fin and drain, then make almond milk, blend the almond in a blending machine, add some water if needed, pass sieve cloth. 3. Add the milk together with almond milk in a pot, mix well and heat until boil. 4. Season with salt, sugar and ground white pepper, then thick the almond milk mixture by using glutinous rice flour and corn starch, pour into a bowl. 5. Add the shark fin, cover steam for 1 hour. Serve immediately.

手磨杏汁炖荷包翅

准备时间：10 分钟。食材：杏仁 100 克，荷包翅 150 克。调料：盐 4 克，白砂糖 6 克，白胡椒粉 2 克，牛奶 20 毫升，纯净水 300 毫升，糯米粉 3 克，鹰粟粉 2 克。制作时间：10 小时。做法：1．将杏仁和荷包翅分别用纯净水浸泡 8 小时。2．把浸泡好的荷包翅冲洗干净。将杏仁与纯净水放入搅拌机中打成浆后过滤备用。3．将过滤好的杏仁浆烧开，加入牛奶，再用糯米粉和鹰粟粉勾浓，并放入盐、白砂糖及白胡椒粉调味。4．把步骤 3 倒入炖盅，放入荷包翅。5．最后把步骤 4 隔水蒸 1 小时即可。

Color Combination　色彩搭配：

THE 65TH PHENOLOGICAL PERIOD
Deer antlers shed

Sensing the resurgence of Yang energy, deer shed their antlers.

65 / 72 物候
麋角解

冬至一阳生，麋角感到阳气萌发，鹿角脱落。

杯深君莫诉　　醉袖歌金缕

宋 · 蔡伸《菩萨蛮》

Caramelized Iced Persimmon
Preparation : 5 mins. Ingredients : persimmons 500g. Seasonings : sugar 20g, shredded white chocolate 3g. Finishing: 3 hrs 30 mins. Method :1. Remove flesh from perssimons and freeze for 3 hours. 2. Sprinkle with sugar and torch to glaze until golden, garnish with shredded white chocolate.

焦糖冻柿子
准备时间：5 分钟。食材：柿子 500 克。调料：白砂糖 20 克，白巧克力碎 3 克。制作时间：3 小时 30 分钟。做法：1. 柿子取肉装入器皿中，放入冰箱冷冻 3 小时。2. 在冻好的柿子上撒上白砂糖，用喷枪燎成金黄色，撒上白巧克力碎即可。

Color Combination　色彩搭配:

横眉群山千秋雪　笑吟长空万里风

唐 · 白居易《涧底松》

Dadong Sea Cucumber With Nostoc

Preparation : 15 mins. Ingredients : sea cucumbers 120g, nostoc 30g, pineapple 10g (scoop 1cm ball) , water 30ml. Seasonings : cooking wine 3ml, dark soy sauce 2ml, light soy sauce 5ml, sugar 1g, chicken stock 100ml, leeks 50g. Finishing : 50 mins. Method : 1. Steam nostoc with water until cooked. 2. Saute leeks in little oil until fragrant then add cooking wine and remaining seasonings, bring to a boil. 3. Add sea cucumbers in and let simmer until tender，cook then arrange on plate, garnish with nostoc and pineapple.

董氏葱烧海参配葛仙米

准备时间: 15 分钟。食材: 海参 120 克, 葛仙米 30 克, 凤梨 10 克（挖成直径约 1 厘米的小球），水 30 毫升。调料: 料酒 3 毫升, 老抽 2 毫升, 酱油 5 毫升, 白砂糖 1 克, 鸡汤 100 毫升, 大葱 50 克 。制作时间: 50 分钟。做法: 1. 将葛仙米与水一起蒸熟备用。2. 起锅放入少量底油, 将大葱煸香, 烹入料酒, 加入其余所有调料。3. 将海参放入汤中, 小火烧至入味收汁, 上盘后点缀以葛仙米及凤梨球。

Color Combination　色彩搭配:

66 / 72 物候 水泉动

THE 66TH PHENOLOGICAL PERIOD
Springs start to flow

阳气萌生,泉水开始流动。

With the rise of Yang energy, spring waters begin to move.

落红片片浑如雾
不教更觅桃源路

清·纳兰性德《海棠春·落红片片浑如雾》

Braised Shark Fin With Saffron Sauce & 30 Years Balsamic Vinegar

Preparation : 3 days. Ingredients : shark fin 300g (soaked). Seasonings : chicken broth 800ml, 30 years balsamic vinegar 3ml, salt 3g, sugar 4g, saffron 1g, chicken fat 10ml, Chinese ham stock 50ml, dried scallop stock 50ml, water starch 10ml. Finishing : 2 hrs. Method : 1.Steam shark fin with 300ml chicken stock for 90 mins. 2.Heat remaining chicken stock with salt, saffron, sugar, chicken fat, Chinese ham stock and dried scallop stock. Then thicken with water starch. 3.Plate the steamed shark fin, dress the saffron sauce. 4.Drop the balsamic vinegar when it serves.

红花汁荷包翅佐意大利 30 年香脂醋

准备时间：3 天。食材：荷包翅 300 克（泡发）。调料：清鸡汤 800 毫升，意大利 30 年香脂醋 3 毫升，盐 3 克，白砂糖 4 克，藏红花 1 克，鸡油 10 毫升，火腿水 50 毫升，干贝水 50 毫升，淀粉水 10 毫升。制作时间：2 小时。做法：1. 将荷包翅与 300 毫升清鸡汤蒸制 90 分钟，备用。2. 锅中加入剩余的清鸡汤、藏红花、盐、白砂糖、鸡油、淀粉水、火腿水及干贝水，煮开后小火苟芡备用。3. 将蒸好的鱼翅浇上红花汁后装盘。4. 最后滴入香脂醋即可。

Color Combination　色彩搭配：

动处清风披拂　展时明月团圆

元·沈禧《风入松 咏扇》

Beijing Sauerkraut With Mini Dired Sausage
Preparation : 10 mins. Ingredients : mini dried sausage 300g, sauerkraut 100g, mud bean vermicelli 100g. Seasonings : chicken stock 800ml, salt 4g, ground white pepper 3g, white vinegar 2ml, lard 10g. Finishing : 20 mins. Method : 1. Fried sauerkraut with lard, add chicken stock to boil, and add remaining seasonings. 2. Add mini dried sausage and cook for 5mins, then add mud bean vermicelli cook for another 3mins to serve.

北京酸菜爆浆东莞肉蛋
准备时间：10 分钟。食材：爆浆肉蛋 300 克，酸菜 100 克，龙口粉丝 100 克。调料：鸡汤 800 毫升，盐 4 克，白胡椒粉 3 克，白醋 2 毫升，猪油 10 克。制作时间：20 分钟。做法：1. 酸菜用猪油煸炒，倒入鸡汤后再放入剩余调料煮开。2. 放入爆浆肉蛋煮 5 分钟，再放入龙口粉丝煮 3 分钟即可。

Color Combination　色彩搭配：

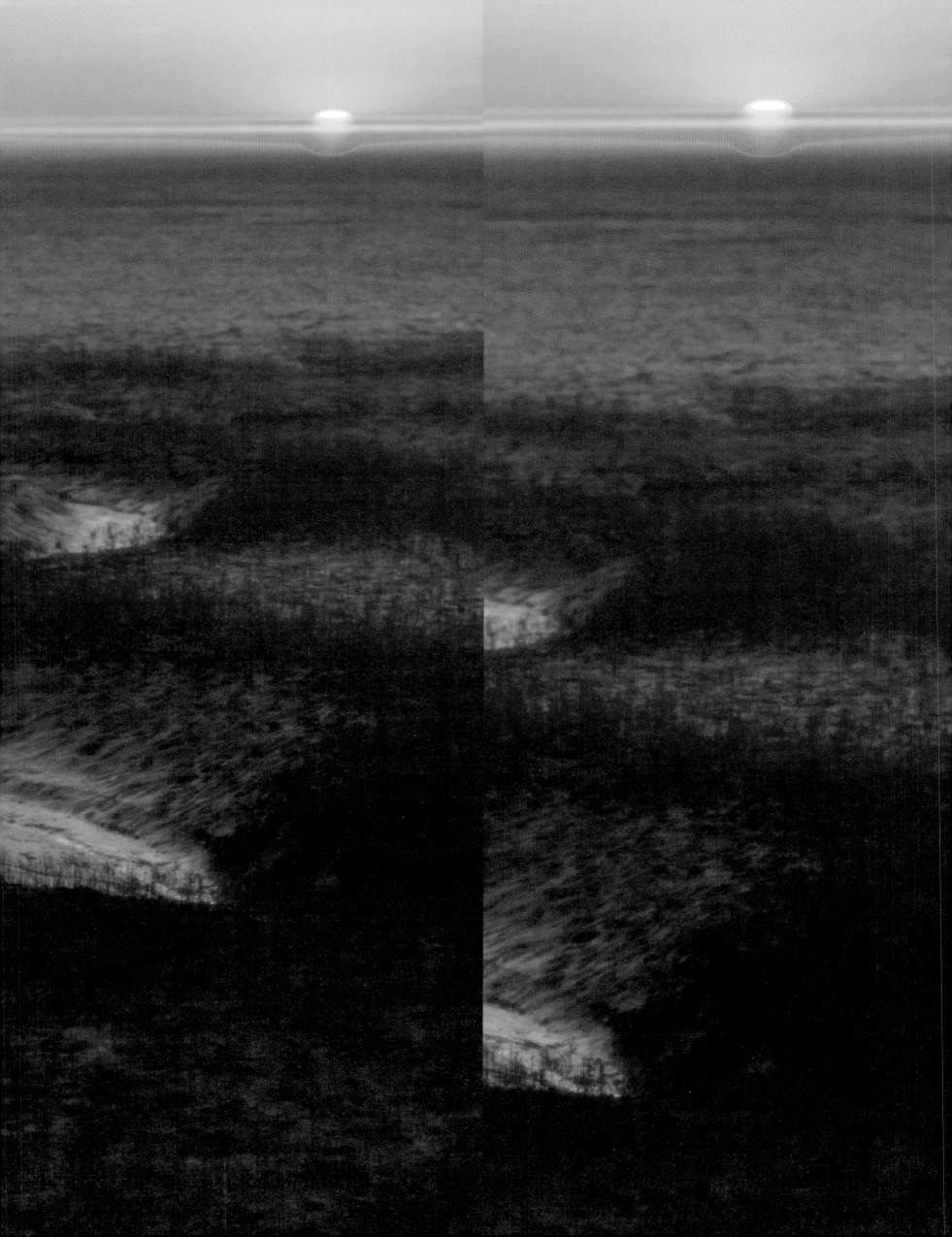

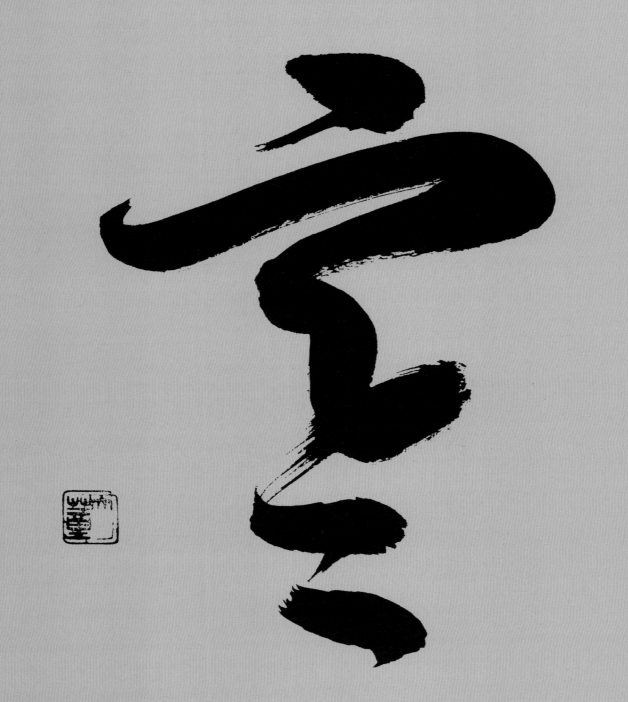

把萝卜头切下来，放碗里，浇上水，端在窗台上，慢慢长出萝卜缨。
冬天没有绿色，萝卜缨成了老百姓家里的景儿。
萝卜缨每天扒着窗户，迎着阳光，自自在在地开花。
萝卜缨开花，开黄花，有淡淡的萝卜香，摘一朵插头上，鲜亮美人。

——《冬吃萝卜夏吃姜》

Cut off the top of the radish, place it in a bowl, pour in some water, and set it by the window. Over time, radish sprouts will grow. In winter, when greenery is a scarcity, these radish sprouts become a delightful sight in many households. Every day, they reach out towards the sunlight, blossoming in their own time. When they bloom, they bear yellow flowers with a faint radish aroma. Pluck a flower, pin it in your hair, and you become a radiant beauty.

— **Eat Radishes in Winter and Ginger in Summer**

小寒

Lesser Cold

THE 67TH PHENOLOGICAL PERIOD
Geese head north

Geese begin their migration north, following the increasing Yang energy.

67 / 72 物候
雁北乡

大雁顺阴阳而迁移,此时阳气已动,大雁开始向北迁移。

荷尽已无擎雨盖　　菊残犹有傲霜枝

宋 · 苏轼《赠刘景文》

Cuttlefish Eggs Soup

Preparation : 20 mins. Ingredients : clean chicken stock 500ml, cuttlefish eggs 150g(soaked). Seasonings : salt 1g, ground white pepper 2g, Shaoxing wine 3ml, water starch 30ml, lecithin 1g, rice vinegar 1ml, sesame oil 1ml, crushed black pepper 1g. Finishing : 30 mins. Method : 1. Blend 100ml clean chicken stock with lecithin, blending until it becomes foam. 2. Pull cuttlefish eggs in thin pieces, then blanch in boiling water with Shaoxing wine, drain. 3. Cook the cuttlefish eggs with remaining clean chicken stock, then add remaining seasonings. 4. Pour the cuttlefish egg soup into coffee cup, top with the foam and garnish with crush black pepper.

泡沫乌鱼蛋汤

准备时间:20分钟。食材:清鸡汤500毫升,乌鱼蛋150克(泡发)。调料:盐1克,白胡椒粉2克,花雕酒3毫升,水淀粉30毫升,大豆卵磷脂1克,米醋1毫升,香油1毫升,黑胡椒碎1克。制作时间:30分钟。做法:1. 将100毫升清鸡汤加热后放入大豆软磷脂,用高速搅拌器打出泡沫备用。2. 乌鱼蛋撕开成片,与花雕酒一起汆水,沥干备用。3. 将剩下的清鸡汤煮开后放入乌鱼蛋,放入其余调料后勾薄芡。4. 把煮好的乌鱼蛋汤倒入咖啡杯中,覆盖上清鸡汤泡沫,撒上黑胡椒碎即可。

Color Combination　色彩搭配:

谁为君种几千枚
小果红墙影里栽

清·袁枚《山楂》

Hawthorn Skewer

Preparation : 5 mins. Ingredients : hawthorn 300g. Seasonings : sugar 500g, water 200ml. Finishing : 40 mins. Method : 1. Skewer the hawthorn. 2. Mix sugar and water, boil until bubbles become small. 3. Dip the hawthorn skewers in and out and set aside until cool.

一串串的糖葫芦

准备时间：5分钟。食材：山楂300克。调料：白砂糖500克，水200毫升。制作时间：40分钟。做法：1.山楂去籽，穿成串。2.白砂糖加水烧开后转小火熬制，直至泡沫变小。3.将山楂串蘸入糖浆，待自然冷却即可。

Color Combination　色彩搭配：

THE 68TH PHENOLOGICAL PERIOD
Magpies start nesting

Magpies, feeling the Yang energy, start building nests for their offspring.

68 / 72 物候
鹊始巢

喜鹊也感受到阳气，开始筑巢，准备繁殖后代。

清似水沉香　色染蔷薇露

宋 · 向子諲《生查子 · 天上得灵根》

Fried Rice

Preparation : 5 mins. Ingredients : fried rice 200g, clean chicken stock 200ml. Seasonings : mustard seeds 3g, sour cream 100ml, lemon zest 3g, lecithin 5g. Finishing : 15 mins. Method : 1. Mix clean chicken stock with lecithin, then blend until become foaming. 2. Place fried rice and sour cream in small cup, then top with chicken stock foam. 3. Garnish with mustard seeds and lemon zest.

合肥炒米泡沫

准备时间：5分钟。食材：炒米200克，清鸡汤200毫升。调料：芥末籽3克，酸奶油100毫升，柠檬皮3克，大豆卵磷脂5克。制作时间：15分钟。做法：1.清鸡汤中加入大豆软磷脂，用搅拌器打发出泡沫备用。 2.将炒米放入小杯中，然后放入酸奶油再盖上泡沫。 3.最后以柠檬皮及芥末籽点缀即可。

Color Combination　色彩搭配：

响松风于蟹眼　浮雪花于兔毫

宋·苏轼《老饕赋》

Alaska King Crab Hot Pot

Preparation : 60 mins. Ingredients : Alaska king crab 1pc (2.5kg). Hot pot stock base: mineral water 1000ml, kelp 5g, dried shrimps 5g, leek 10g, ginger 10g. Seasoning dressing: sesame paste 50g, soy sauce for seafood 50ml, curry sauce 50ml, spicy Puning bean sauce 50g. Finishing : 10 mins. Method : 1. Chop Alaska king crab in pieces, then arrange on plate. 2. Boil hot pot stock base. 3. Serve with 4 kinds of seasoning dressing.

涮阿拉斯加蟹

准备时间：60 分钟。食材：阿拉斯加蟹 1 只（约 2.5 公斤）。汤底：矿泉水 1000 毫升，昆布 5 克，海米 5 克，葱段 10 克，姜片 10 克。蘸料：麻酱 50 克，海鲜酱油 50 毫升，咖喱汁 50 毫升，辣普宁豆瓣酱 50 克。制作时间：10 分钟。做法：1. 阿拉斯加蟹切块，码放在盘中备用。2. 将汤底原材料一起煮开。3. 食用时搭配蘸料即可。

Spicy Puning Bean Sauce / Ingredients : puning bean paste 350g, fried garlic 100g, fried shallot 100g, fried dried scallop 50g, ground peanut 50g, ground white sesame 50g, sugar 60, sesame oil 100g, chili oil 50ml, Hunan chili paste 70g, soup-stock 300ml, fried dried shrimps 50g, vegetable oil 100g. Method: 1. Sweat Hunan chili paste with vegetable oil, add fried garlic, fried shallot, fried dried scallop and fried dried shrimps. 2. Add Puning bean paste and soup-stock stir-fry in low heat for 10mins. 3. Add ground peanut, ground white sesame and sugar, fry for another 10mins, then add sesame oil and chili oil.

辣普宁豆酱 / 食材：普宁豆酱 350 克，炸蒜蓉 100 克，炸红葱头 100 克，炸干贝丝 50 克，花生粉 50 克，白芝麻粉 50 克，白砂糖 60 克，香油 100 毫升，辣油 50 毫升，湖南辣椒酱 70 克，清汤 300 毫升，炸海米碎 50 克，色拉油 100 毫升。做法：1. 锅中放入色拉油，将湖南辣椒酱煸香，加入炸红葱头、炸干贝丝、炸海米碎、炸蒜蓉。2. 放入普宁豆酱，加入清汤，小火炒 10 分钟。3. 加入花生粉、白芝麻粉、白砂糖，再以小火炒 10 分钟后加入香油和辣油，搅匀即可。

Curry Sauce / Ingredients : onion 80g (finely chopped), chili 200g (finely chopped), celery 200g (finely chopped), shallot 200g (finely chopped), curry powder 150g, coconut cream 2.4l, evaporated filled milk 820ml, soup-stock 500ml, salt 8g, sugar 6g, fish sauce 20ml, white roux 160g, butter 180g. Method: 1. Sweat onion, chili, celery and shallot by butter, add soup-stock to boil. 2. Take away onion, chili, celery and shallot, add curry powder cook for 5mins. 3. Add coconut cream, evaporated filled milk, salt, sugar and fish sauce to taste, then stir in white roux for thickening.

咖喱酱 / 食材：洋葱 80 克（切碎），尖椒 200 克（切碎），西芹 200 克（切碎），红葱 200 克，咖喱粉 150 克，椰浆 2.4 升，三花淡奶 820 毫升，清鸡汤 500 毫升，盐 8 克，白砂糖 6 克，鱼露 20 毫升，黄油炒面 160 克，黄油 180 克。做法：1. 先将洋葱、尖椒、西芹、红葱用黄油煸香，然后加入清鸡汤烧开。2. 将步骤 1 中的洋葱、尖椒、西芹及红葱捞出，再放入咖喱粉，小火熬煮 5 分钟。3. 加入椰浆、三花淡奶，搅匀后加入盐、白砂糖、鱼露调味，最后放入黄油炒面，拌匀即可。

Color Combination　色彩搭配：

THE 69TH
PHENOLOGICAL PERIOD
Wild chickens crow for mates

With the growth of Yang energy,
wild chickens crow in search of mates.

69 / 72 物候
雉始雊

阳气滋长，野鸡开始鸣叫求偶。

Steam Taro With Dark Sugar
Preparation : 5 mins. Ingredients : taro 180g. Seasonings : dark sugar 30g. Finishing : 50 mins. Method : 1. Cut taro 2cmX2cmX5cm pieces, steam for 30mins. 2. Bake dark sugar in 85℃ for 10mins, then rub into fine powder. 3. Dress dark sugar on the steamed taro to serve.

赤糖福鼎槟榔芋头
准备时间：5分钟。食材：芋头180克。调料：红糖30克。制作时间：50分钟。做法：1. 将芋头切成长2厘米、宽2厘米、高5厘米的块，蒸30分钟，备用。2. 将红糖放入预热至85℃的烤箱中烤10分钟，搓碎。3. 将红糖撒在芋头上即可。

玉脂如肪粉且柔
芋魁芋魁满载瓯

明·屠本畯《蹲鸱》

Color Combination 色彩搭配：

冰盘荐琥珀　何似糖霜美

宋 · 苏轼《送金山乡僧归蜀开堂》

Egg Caramel Treats

Preparation : 10 mins. Ingredients : egg yolk 250g, egg 1pc, baking powder 3g, flour 500g. Seasonings : water 180ml, sugar 100g, glucose 500g. Finishing : 3 hrs. Method : 1. Mix egg yolk, egg, baking powder and flour into dough, then roll into 3mm thin, slice into thin slices, rest for 1hr. 2. Deep fry the dough until golden colour. 3. Boil the remaining seasonings into thick syrup, mix with the fried dough, then press into mould and let cool to set. 4. Slice the dough into square pieces, then sevre.

葵香鸡蛋萨其马

准备时间：10 分钟。食材：葵香鸡蛋黄 250 克，葵香鸡蛋 1 只，泡打粉 3 克，面粉 500 克。调料：水 180 毫升，白砂糖 100 克，葡萄糖 500 克。制作时间：3 小时。做法：1. 将葵香鸡蛋黄、葵香鸡蛋、面粉、泡打粉和成面团，擀成 3 毫米左右的薄片后切成丝，醒发 1 小时。2. 将面丝放入油锅中炸至金黄色，沥干油。3. 葡萄糖、白砂糖与水熬成糖浆，与炸好的面丝拌匀，再放入模具摊凉成型。4. 食用时切成正方形小块即可上盘。

富平柿饼是大自然的馈赠。最好的柿饼在最冷的冬天。除了空口吃，柿饼还能切开，放冰淇淋或者奶酪，我自己这样吃。

——《吃富平柿饼，过如意春节》

The dried persimmon from Fuping, Shaanxi province is nature's generous offering. Best savored in the heart of winter, they can be enjoyed in their natural state or given a modern twist with a scoop of ice cream or a slice of cheese. That's the way I relish them.

— **Savoring Fuping Dried Persimmon, Celebrating a Joyful Spring Festival**

大寒

Greater Cold

THE 70TH PHENOLOGICAL PERIOD
Chickens begin to nurture

Chickens start to take care of their offspring.

70 / 72 物候
鸡始乳

■ 鸡开始哺育后代。

Steak Rossini With Caramel Foie Gras
Preparation : 20 mins. Ingredients : dry aged sirloin 180g, foie gras 120g (sliced), orange 4pcs (fillet), mashed potato 70g. Seasonings : ground white pepper 3g, cognac 2ml, sea salt 3g, caramel 10g, butter 25g, crushed black pepper 5g. Finishing : 30 mins. Method : 1. Pan fry orange fillet until caramelized, heat butter and mashed potato. Set aside and keep warm. 2. Marinade foie gras with salt and ground white pepper, cognac for 10mins, then pan fry both side, dress with caramel. 3. Pan fry DA sirloin for 3mins. 4.Arrange orange fillet and mashed potato on plate. 5. Garnish with crushed black pepper and dress with hot butter.

■

干式熟成罗西尼牛排和焦糖肥肝
准备时间：20 分钟。食材：熟成西冷牛排 180 克，鹅肝（切片）120 克，橙子（切角）4 片，土豆泥 70 克。调料：白胡椒粉 3 克，干邑 2 毫升，海盐 3 克，焦糖 10 克，黄油 25 克，黑胡椒碎 5 克。制作时间：30 分钟。做法：1. 橙子角用煎锅煎出焦面，土豆泥加黄油炒制，保温备用。2. 鹅肝用海盐、白胡椒粉、干邑腌制 10 分钟，煎至两面呈金黄色，淋上焦糖。3. 牛排两面煎 3 分钟，备用。4. 盘中放入橙子角及土豆泥，再放上牛排和煎鹅肝。5. 最后撒上黑胡椒碎，浇上热黄油即可。

芳莲坠粉
疏桐吹绿
庭院暗雨乍歇

宋·姜夔《八归·湘中送胡德华》

Color Combination 色彩搭配：

小山重叠金明灭
鬓云欲度香腮雪

唐·温庭筠《菩萨蛮·小山重叠金明灭》

Da Dong's Ice-Cream

Preparation : 5 mins. Ingredients : dark chocolate 300g, cream cheese 500g, vanilla ice-cream 200g. Seasonings : rose jam 20g, passion fruit jam 30g, dark coco-powder 1g. Finishing : 7 hrs. Method : 1. Pipe vanilla ice-cream into silicon mould, put in wood stick then put in deep freezer for 6 hours. 2. Melt dark chocolate by bain marie, then dip the frozen vanilla ice-cream into make chocolate shell, back to deep freezer. 3. Mix cream cheese with rose jam and passion fruit jam to make 2 kinds of cheese ball. 4. Dust the dark coco-powder on plate, arrange ice-cream and garnish with cheese ball.

大董雪糕

准备时间：5 分钟。食材：黑巧克力 300 克，奶油奶酪 500 克，香草冰淇淋 200 克。调料：玫瑰花馅 20 克，热情果酱 30 克，黑可可粉 1 克。制作时间：7 小时。做法：1. 将香草冰淇淋挤入硅胶磨具，插上木条后冷冻 6 小时。 2. 黑巧克力隔水融化，将冷冻的冰淇淋取出，蘸入黑巧克力中裹上外壳，再冷冻保存。 3. 将奶油奶酪分别与玫瑰花馅及热情果酱混合成奶酪球。 4. 把黑可可粉撒在盘上，再放上冰棍雪糕及奶酪球即可。

Color Combination 色彩搭配：

THE 71ST PHENOLOGICAL PERIOD
Predatory birds become fierce

In the deep cold, raptors hunt more aggressively, seizing food to ward off the chill.

71 / 72 物候
征鸟厉疾

大寒时节天气更加寒冷，鹰隼捕食变得凶狠快速，强悍地抢夺更多食物以抵御寒冷。

燕子双飞来又去　　纱窗几度春光暮

宋 · 苏轼《蝶恋花 · 记得画屏初会遇》

L'Oeuf Au Crab

Preparation : 10 mins. Ingredients : mandarin fish fillet 250g, crab roe 150g, lysimachia christinae 3g, carrot tops 2g, honey pea 2g (blanched). Seasonings : pork lard 10g, salt 5g, soy sauce 3ml, ginger juice 10ml, water of ginger and spring onion 300ml, basil oil 2ml. Finishing : 8 hrs. Method : 1. Soak mandarin fish fillet in ice water for 6 hours, drain. 2. Put the fillet together with water of ginger and spring onion into food processor, then blend into mousse. 3. Fried crab roe with pork lard, then season with ginger juice, soy sauce and salt. 4. Freeze the crab roe, then use baller to form into small ball. 5. Fill the fish mousse into piping bag, then fill half full into an egg-shell. 6. Stuff the crab roe ball into the egg-shell, then fill up by fish mousseline. 7. Put the egg-shell into 80° C degree water, then poach for 12 mins. 8. Drain the egg-shell, break the egg shell then arrange on plate. 9. Garnish lysimachia christinae, carrot tops, honey pea and basil oil when time serve.

灌蟹芙蓉蛋

准备时间：10分钟。食材：鳜鱼鱼肉250克，蟹粉150克，金钱草3克，胡萝卜缨2克，小嫩豆2克（汆水）。调料：猪油10克，盐5克，鲜酱油3毫升，姜汁10毫升，葱姜水300毫升，罗勒油2毫升。制作时间：8小时。做法：1. 鳜鱼鱼肉用冰水浸泡6小时，与葱姜水一起放入搅拌机中打成鱼蓉，再放入冰箱冷藏备用。2. 蟹粉用猪油炒热，再用姜汁、盐、鲜酱油调味。3. 把步骤2放入冰箱冰冻后，用球勺挖成球状。4. 将步骤1装入挤袋，在蛋壳中挤入一半的量，放入蟹粉球，之后再用鱼蓉填满。5. 将步骤4放入80°C的水中煮12分钟。6. 将鸡蛋壳打开上盘，再配以金钱草、胡萝卜缨、小嫩豆，淋上罗勒油即可。

沙鸥径去鱼儿饱　野鸟相呼柿子红

宋 · 郑刚中《晚望有感》

Persimmon Salad

Preparation : 10 mins. Ingredients : 1 persimmon, persimmon juice 200ml. Seasonings : olive oil 3ml, lemon juice 1ml. Finishing : 20 mins. Method : 1. Wash the persimmon, then mold flesh into ball. Toss persimmon juice with liquid nitrogen into sorbet. 2. Season persimmon ball with olive oil and lemon juice. 3. Place persimmon in serving bowl and garnish with persimmon sorbet.

脆柿子沙拉

准备时间：10 分钟。食材：脆柿子 1 个，柿子汁 200 毫升 。调料：橄榄油 3 毫升，柠檬汁 1 毫升 。制作时间：20 分钟。做法：1. 柿子洗净去蒂，用挖球器将柿子肉挖成球状。柿子汁则用液氮炒成冰沙。2. 取出挖好的柿子球，加橄榄油、柠檬汁拌匀。3. 将拌好的柿子球与柿子冰沙一起装盘即可。.

THE 72ND PHENOLOGICAL PERIOD
Ice becomes solid in the center of lakes

With Yang energy still weak and the east wind yet to arrive,
ice thickens to the center of lakes, making them sturdy and solid.

72 / 72 物候
水泽腹坚

阳气未达，东风未至，湖面上的冰会结到湖中央，整个冰面变得非常坚固。

霜降红梨熟
柔柯已不胜

宋·苏轼《梨》

Hawthorn Cake With Shredded Pears

Preparation : 5 mins. Ingredients : pears 300g, hawthorn cake 100g. Finishing : 5 mins. Method : 1. Clean and shred pears. 2. Scoop balls from hawthorn cake and plate with shred pears.

京糕梨丝

准备时间：5 分钟。食材：梨 300 克，京糕 100 克。
制作时间：5 分钟。做法：1. 梨洗净，切丝。2. 将京糕用挖球器挖成球状，与梨丝装盘即可。

多情只有春庭月　犹为离人照落花

唐 · 张泌《寄人》

Petit Fours

小吃八款

Steamed Rice Cakes With Nut Stuffing

Preparation : 3hrs. Ingredients : glutinous rice 500g, flour 300g(steamed). Seasonings : walnuts 100g(toasted), pumpkin seed 100g(toasted), hawthorn jelly 100g. Finishing : 60 mins. Method : 1. Soak glutinous rice for 3 hours, then steam for 30mins. 2. Mix seasonings into stuffing. 3. Mix steamed glutinous rice with flour, then wrap the stuffing, and form into ball shape.

艾窝窝

准备时间：3 小时。食材：糯米 500 克，面粉 300 克（蒸熟）。调料：核桃仁 100 克，瓜子仁 100 克，山楂糕 100 克。制作时间：60 分钟。做法：1. 将糯米浸泡 3 小时后隔水蒸 30 分钟。 2. 核桃仁、瓜子仁、山楂糕拌匀成馅。 3. 将蒸好的糯米与面粉拌匀，包裹上果仁馅即成。

Glutinous Rice Rolls With Sweet Bean

Preparation : 15 mins. Ingredients : glutinous rice flour 250g, cream 50ml, rice flour 25g, water 300ml.Seasonings: ground soy bean flour 100g, red bean paste 80g. Finishing : 50 mins. Method : 1.Mix glutinous rice flour, cream, rice flour and water into dough, then steam for 30mins. 2.Dust ground soy bean flour on broad, roll the dough into 3mm , spread rea bean paste then roll. 3.Slice into short pieces when it serves.

驴打滚

准备时间：15 分钟。食材：糯米粉 250 克，奶油 50 毫升，黏米粉 25 克，水 300 毫升。调料：黄豆面 100 克，豆沙 80 克。制作时间：50 分钟。做法：1. 混合糯米粉、黏米粉、水、奶油，隔水蒸 30 分钟。2. 案板上撒一层黄豆面，将面团擀成 3 毫米的薄皮，再抹一层豆沙后卷起。 3. 食用时将卷切段即可。

Color Combination　色彩搭配：

390

Steamed Mini Corn Bread

Preparation : 15 mins. Ingredients : corn flour 400g, sugar 50g, baking powder 5g, yeast 5g, instant soy milk powder 100g, water 300ml. Finishing : 45 mins. Method : 1. Mix all ingredients into dough. 2. Form the dough into 30g mini cone shape each, then steam for 30mins.

小窝头

准备时间：15 分钟。食材：玉米面 400 克，白砂糖 50 克，泡打粉 5 克，酵母 5 克，豆浆粉 100 克，水 300 毫升。制作时间：45 分钟。做法：1. 玉米面、豆浆粉、白砂糖、泡打粉、酵母和水拌匀成面团。2. 将面团捏成窝头形状，隔水蒸 30 分钟即可。

White French Bean Roll

Preparation : 5 mins. Ingredients : white French bean 500g. Seasonings : red bean paste 30g, white sesame 30g (toasted). Finishing : 60 mins. Method : 1. Steam white French bean for 30mins, then pass sieve into fine paste. 2. Mix red bean paste and white sesame. 3. Place cling wrap on bamboo sushi map, spread white french bean paste, then pipe the red bean paste mixture, roll up. 4. Cut into small pieces while it serves.

芸豆卷

准备时间：5 分钟。食材：芸豆 500 克。调料：红豆沙 30 克，白芝麻 30 克。制作时间：60 分钟。做法：1. 芸豆去皮蒸 30 分钟，压过细筛成芸豆沙备用。2. 红豆沙和白芝麻混合。3. 寿司帘上垫上保鲜膜，放上芸豆沙并压成片，再挤上芝麻红豆沙，再卷成卷。4. 最后切小块即可。

Fried Crispy Slice, Sweet Ginger Jus

Preparation : 5 mins. Ingredients : flour 500g. Seasonings : salt 5g, ginger juice 10ml, syrup 200ml, water 200ml. Finishing : 40 mins. Method : 1. Mix flour, salt and water into dough, then roll into 2mm thin. Deep fry till golden color, drain. 2. Heat ginger juice and syrup, until thick. 3. Dress ginger syrup on the fried slices when it serves.

姜汁排叉

准备时间：5 分钟。食材：面粉 500 克。调料：盐 5 克，姜汁 10 毫升，糖水 200 毫升，水 200 毫升。制作时间：40 分钟。做法：1. 面粉加盐和水揉成面团，擀成 2 毫米薄片并翻出花型，油炸至金黄色，沥干油。2. 加热糖水与姜汁，慢火熬成糖浆。3. 最后将姜汁糖水淋在步骤 1 上即可。

Pea Purée Cake

Preparation : 1hr. Ingredients : skinless sweet peas 500g, water 1l. Seasonings : xylitol 200g, agar 30g(soaked). Finishing : 2hrs. Method : 1. Steam sweet peas with water for 1hr, then pass sieve into fine paste. 2. Mix remaining seasonings with sweet peas purée with low heat until thick, then move into square mould. Let cool. 3. Slices while it serves.

豌豆黄

准备时间：1 小时。食材：去皮豌豆 500 克，水 1 升。调料：木糖醇 200 克，琼脂 30 克（浸泡）。制作时间：2 小时。做法：1. 豌豆与水煮 1 小时至软烂，过筛成豌豆泥。2. 豌豆泥中加入木糖醇及琼脂烧开，待熬至黏稠倒入模具，晾凉。3. 食用时切成小块即可。

Fried Mini Sweet Dough

Preparation : 5 mins. Ingredients : levain 500g, flour 200g, alkali powder 3g, baking powder 5g. Seasonings : maltose 700g, dried osmanthus 10g. Finishing : 1 hr. Method : 1. Mix 500g maltose and dried osmanthus then steam for 20mins. 2. Mix levain, alkali powder and baking powder, then roll thin. 3. Mix flour with remaining maltose. 4. Layer the maltose flour on the levain skin, then fold over. 5. Slice in small piece, deep fry until golden color. 6. Soak fried dough into osmanthus syrup.

糖耳朵

准备时间：5 分钟。食材：老面肥 500 克，面粉 200 克，碱面 3 克，泡打粉 5 克。调料：饴糖 700 克，干桂花 10 克。制作时间：1 小时。做法：1. 500 克饴糖加干桂花蒸 20 分钟备用。2. 老面肥加碱面和泡打粉后和面。3. 面粉中加入剩余饴糖和成糖心。4. 面团分成两份，中间浇入一层糖心，擀开后切成合适的大小，炸至金黄。5. 把炸好的步骤 3 泡在饴糖里即可。

Fried Chinese Yam Rolls

Preparation : 10 mins. Ingredients : mashed Chinese yam 250g, oil soy bean cure skin sheet 5pcs, carrot 1000g (finely sliced), sliced sugar pickle mandarin's skin 100g, green plum meat 50g (finely sliced), flour 80g. Finishing : 60 mins. Method :1. Mix mashed Chinese yam, carrot slice, sliced sugar pickle mandarins skin, sliced green plum meat and flour into stuffing. 2. Spread the stuffing on the oil soy bean cure skin sheet then roll, steam for 30mins. 3. Until cool, slice in piece. 4. Deep fry while it serves.

糖卷果

准备时间：10 分钟。食材：山药泥 250 克，油豆皮 5 张，胡萝卜 1000 克（切丝），红丝 100 克，青梅肉 50 克（切丝），面粉 80 克。制作时间：60 分钟。做法：1. 分别把胡萝卜丝、红丝、青梅肉、山药泥与面粉拌匀。2. 油豆皮包上步骤 1 中的馅料，蒸 30 分钟，摊凉备用。3. 食用时把步骤 2 切块，油炸成金黄即可。

未经许可，不得以任何方式复制或抄袭本书之部分或全部内容。

版权所有，侵权必究。

图书在版编目（CIP）数据

大董中国意境菜. 2023：七十二物候 / 大董著；谢安冰编；（澳）杰夫·龙（Geoff Lung），大董摄影. — 北京：电子工业出版社，2023.12

ISBN 978-7-121-46791-2

Ⅰ. ①大… Ⅱ. ①大… ②谢… ③杰… Ⅲ. ①中式菜肴—菜谱 Ⅳ. ①TS972.182

中国国家版本馆CIP数据核字(2023)第228871号

责任编辑：白　兰
印　　刷：北京雅昌艺术印刷有限公司
装　　订：北京雅昌艺术印刷有限公司
出版发行：电子工业出版社
　　　　　北京市海淀区万寿路173信箱　邮编：100036
开　　本：880×1230　1/8　印张：49.25　字数：745千字
版　　次：2023年12月第1版
印　　次：2023年12月第1次印刷
定　　价：798.00元

凡所购买电子工业出版社图书有缺损问题，请向购买书店调换。若书店售缺，请与本社发行部联系，联系及邮购电话：（010）88254888，88258888。

质量投诉请发邮件至zlts@phei.com.cn，盗版侵权举报请发邮件至dbqq@phei.com.cn。

本书咨询联系方式：bailan@phei.com.cn，（010）68250802。